PRAISE FOR *THE GLASS UNIVERSE*

"An elegant historical tale . . . [from] the master storyteller of astronomy."
—THE BOSTON GLOBE

"A joy to read . . . Sobel writes with an eye for a telling detail and an ear for an elegant turn of phrase."
—THE WALL STREET JOURNAL

"Unforgettable . . . It takes a talented writer to interweave professional achievement with personal insight. By the time I finished *The Glass Universe*, Dava Sobel's wonderful, meticulous account, it had moved me to tears."
—Sue Nelson, NATURE

"Sensitive, exacting, and lit with the wonder of discovery."
—Elizabeth Kolbert, Pulitzer Prize–winning author of THE SIXTH EXTINCTION

"Insightful, uplifting, accessible, compelling and an absolute joy . . . Everyone with even a vague interest in the early history of astrophysics and specifically the huge contribution made by women to its advancement should read this superb, rewarding book."
—THE OBSERVATORY, The Royal Astronomical Society

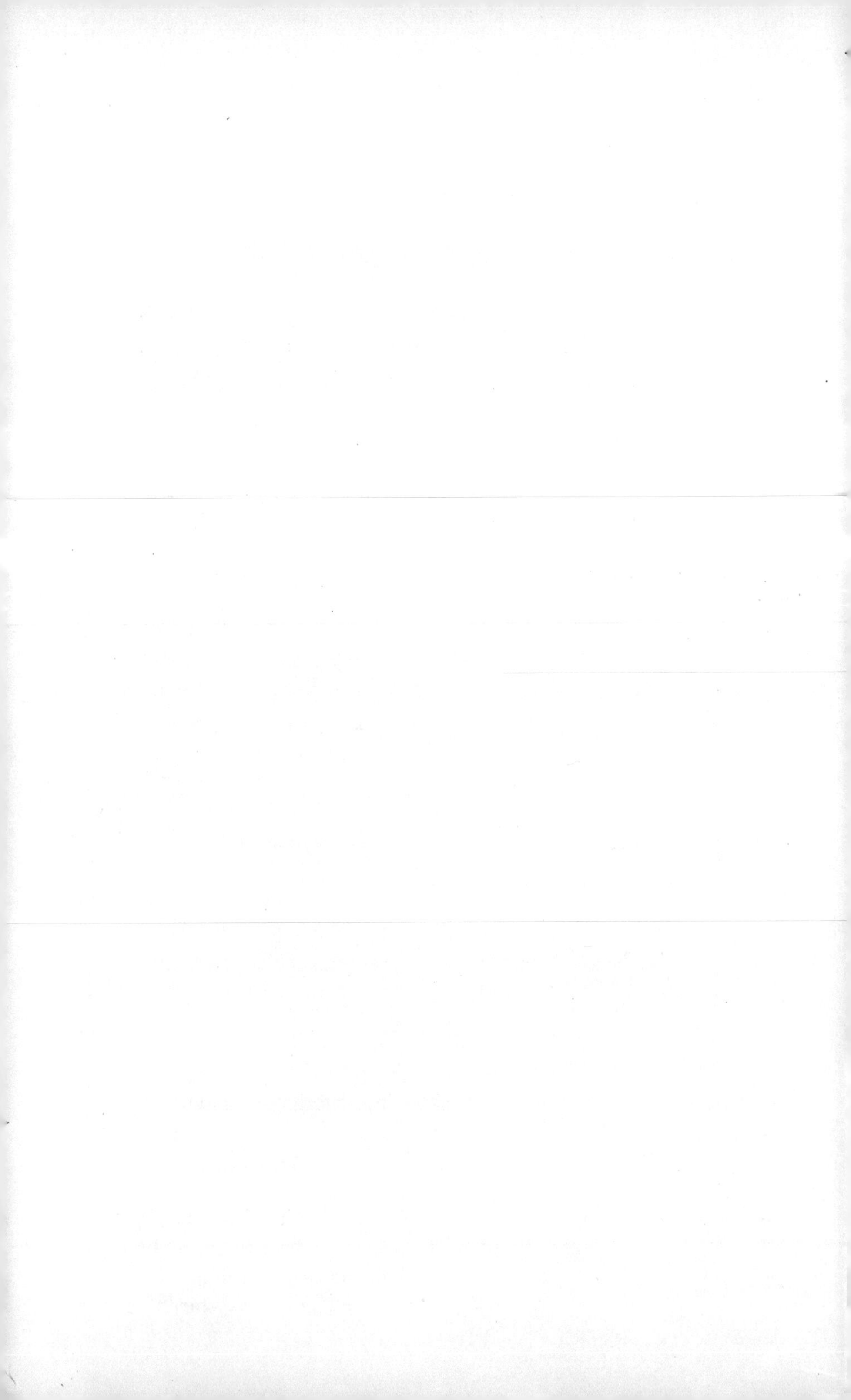

Praise for *The Glass Universe*

Finalist for the 2017 Indies Choice Awards
Longlisted for the PEN/E. O. Wilson Literary Science Writing Award and the Carnegie Medal

"Sobel mixes discussions of the most abstruse topics with telling glimpses of her subjects' lives, in the process showing how scientific and social progress often go hand in hand." —*The New Yorker*

"[Sobel] traces a remarkable line in American female achievement . . . [and] captures the stalwart spirit of Pickering's female finds." —*USA Today*

"A fascinating and inspiring tale of . . . female pioneers who have been shamefully overlooked." —*Real Simple*

"An astronomically large topic generously explored."
—*O, The Oprah Magazine*

"This is intellectual history at its finest."
—Geraldine Brooks, *New York Times* bestselling author of *The Secret Chord*

"Sobel has distinguished herself with lucid books about scientists and their discoveries. . . . [She] vividly captures how her brilliant and ambitious protagonists charted the skies and found personal fulfillment in triumphant discovery." —*The National Book Review*

"A compelling read and a welcome reminder that American women have long desired to reach for the stars." —*BookPage*

"[Sobel] soars higher than ever before . . . [continuing] her streak of luminous science writing with this fascinating, witty, and most elegant history. . . . *The Glass Universe* is a feast for those eager to absorb forgotten stories of resolute American women who expanded human knowledge."
—*Booklist* (starred review)

"Sobel knows how to tell an engaging story. . . . With grace, clarity, and a flair for characterization, [she] places these early women astronomers in the wider historical context of their field for the very first time."
—*Publishers Weekly* (starred review)

PENGUIN BOOKS

THE GLASS UNIVERSE

Dava Sobel is the author of five books, including the *New York Times* bestselling *Longitude*, *Galileo's Daughter*, and *The Planets*. A former *New York Times* science reporter and a longtime contributor to *The New Yorker*, *Audubon*, *Discover*, and *Harvard Magazine*, she is a recipient of the National Science Board's Individual Public Service Award and the Boston Museum of Science's Bradford Washburn Award, as well as numerous literary prizes.

Visit davasobel.com

The GLASS UNIVERSE

HOW THE LADIES *of*
the HARVARD OBSERVATORY
TOOK THE MEASURE *of the* STARS

Dava Sobel

PENGUIN BOOKS

PENGUIN BOOKS
An imprint of Penguin Random House LLC
375 Hudson Street
New York, New York 10014
penguin.com

First published in the United States of America by Viking Penguin,
an imprint of Penguin Random House LLC, 2016
Published in Penguin Books 2017

Insert page 1, bottom: Angelo Secchi, *Le soleil*, 1875–1877.
Insert page 2, top: Courtesy of Carbon County Museum, Rawlins, Wyoming
Insert page 3, bottom: UAV 630.271 (E4116), Harvard University Archives
Insert page 4, top: Courtesy of Harvard College Observatory
Insert page 5, top: Schlesinger Library, Radcliffe Institute, Harvard University; bottom: Courtesy of Hastings Historical Society, New York
Insert page 6, top: Courtesy of Harvard College Observatory; bottom: Lindsay Smith, used with permission
Insert page 7, top: HUGFP 125.82p, Box 2, Harvard University Archives; bottom: Special Collections Research Center, University of Chicago Library
Insert page 8, top: HUPSF Observatory (14), olvwork360662, Harvard University Archives; pages 8–9 bottom: UAV 630.271 (391), olvwork432043, Harvard University Archives
Insert page 9, top: Courtesy of Harvard College Observatory
Insert page 10, top: HUGFP 125.82p, Box 2, Harvard University Archives; bottom: Courtesy of Harvard College Observatory
Insert page 11, top: HUGFP 125.36 F, Box 1, Harvard University Archives; bottom: HUGFP 125.36 F, Box 1, Harvard University Archives
Insert page 12, bottom left and right: Courtesy of Katherine Haramundanis
Insert page 13, top: Courtesy of the Harvard University Archives; bottom: Courtesy of Charles Reynes
Insert page 14: Chart 1, Volume 105, Harvard College Observatory Annals
Insert page 15, top: Courtesy of Hastings Historical Society, New York; bottom: Courtesy of Katherine Haramundanis
Insert page 16, top: Lia Halloran, used with permission; bottom: Richard E. Schmidt, used with permission

ISBN 9780143111344 (paperback)

THE LIBRARY OF CONGRESS HAS CATALOGED THE HARDCOVER EDITION AS FOLLOWS:
Names: Sobel, Dava.
Title: The glass universe : how the ladies of the Harvard Observatory took the measure of the stars / Dava Sobel.
Description: New York : Viking, 2016. | Includes bibliographical references and index.
Identifiers: LCCN 2016029496 (print) | LCCN 2016030208 (e-book) | ISBN 9780670016952 (hardcover) | ISBN 9780698148697 (e-book)
Subjects: LCSH: Women in astronomy—Massachusetts—History. | Women mathematicians—Massachusetts—History. | Astronomy—History—19th century. | Astronomy—History—20th century. | Harvard College Observatory.
Classification: LCC QB34.5 .S63 2016 (print) | LCC QB34.5 (ebook) | DDC 522/.19744409252—dc23
LC record available at https://lccn.loc.gov/2016029496

Printed in the United States of America
3 5 7 9 10 8 6 4

Set in Warnock Pro
Designed by Amy Hill

To the ladies who sustain me:
Diane Ackerman, Jane Allen,
KC Cole, Mary Giaquinto, Sara James, Joanne Julian,
Zoë Klein, Celia Michaels, Lois Morris,
Chiara Peacock, Sarah Pillow,
Rita Reiswig, Lydia Salant, Amanda Sobel,
Margaret Thompson, and Wendy Zomparelli,
with love and thanks

CONTENTS

PREFACE

A LITTLE PIECE OF HEAVEN. That was one way to look at the sheet of glass propped up in front of her. It measured about the same dimensions as a picture frame, eight inches by ten, and no thicker than a windowpane. It was coated on one side with a fine layer of photographic emulsion, which now held several thousand stars fixed in place, like tiny insects trapped in amber. One of the men had stood outside all night, guiding the telescope to capture this image, along with another dozen in the pile of glass plates that awaited her when she reached the observatory at 9 a.m. Warm and dry indoors in her long woolen dress, she threaded her way among the stars. She ascertained their positions on the dome of the sky, gauged their relative brightness, studied their light for changes over time, extracted clues to their chemical content, and occasionally made a discovery that got touted in the press. Seated all around her, another twenty women did the same.

The unique employment opportunity that the Harvard Observatory afforded ladies, beginning in the late nineteenth century, was unusual for a scientific institution, and perhaps even more so in the male bastion of Harvard University. However, the director's farsighted hiring practices, coupled with his commitment to systematically photographing the night sky over a period of decades, created a field for women's work in a glass universe. The funding for these projects came primarily from two heiresses with abiding interests in astronomy, Anna Palmer Draper and Catherine Wolfe Bruce.

The large female staff, sometimes derisively referred to as a harem, consisted of women young and old. They were good at math, or devoted stargazers, or both. Some were alumnae of the newly founded women's colleges, though others brought only a high school education and their own native ability. Even before they won the right to vote, several of them made contributions of such significance that their names gained honored places in the history of astronomy: Williamina Fleming, Antonia Maury, Henrietta Swan Leavitt, Annie Jump Cannon, and Cecilia Payne. This book is their story.

PART ONE

The Colors of Starlight

I swept around for comets about an hour, and then I amused myself with noticing the varieties of color. I wonder that I have so long been insensible to this charm in the skies, the tints of the different stars are so delicate in their variety. . . . What a pity that some of our manufacturers shouldn't be able to steal the secret of dyestuffs from the stars.

—Maria Mitchell (1818–1889)
Professor of Astronomy, Vassar College

The white mares of the moon rush along the sky
Beating their golden hoofs upon the glass heavens

—Amy Lowell (1874–1925)
Winner of the Pulitzer Prize for Poetry

CHAPTER ONE

Mrs. Draper's Intent

THE DRAPER MANSION, uptown on Madison Avenue at Fortieth Street, exuded the new glow of electric light on the festive night of November 15, 1882. The National Academy of Sciences was meeting that week in New York City, and Dr. and Mrs. Henry Draper had invited some forty of its members to dinner. While the usual gaslight illuminated the home's exterior, novel Edison incandescent lamps burned within—some afloat in bowls of water—for the amusement of the guests at table.

Thomas Edison himself sat among them. He had met the Drapers years ago, on a camping trip in the Wyoming Territory to witness the total solar eclipse of July 29, 1878. During that memorable interlude of midday darkness, as Mr. Edison and Dr. Draper executed their planned observations, Mrs. Draper had dutifully called out the seconds of totality (165 in all) for the benefit of the entire expedition party, from *inside* a tent, where she remained secluded, blind to the spectacle, lest the sight of it unnerve her and cause her to lose count.

The red-haired Mrs. Draper, an heiress and a renowned hostess, surveyed her electrified salon with satisfaction. Not even Chester Arthur in the White House lighted his dinner parties with electricity. Nor could the president attract a more impressive assembly of science's luminaries. Here she welcomed the well-known zoologists Alexander Agassiz, down from Cambridge, Massachusetts, and Spencer Baird, up from the Smithsonian

Institution in Washington. She introduced her family friend Whitelaw Reid of the *New York Tribune* to Asaph Hall, world famous for his discovery of Mars's two moons, and to solar expert Samuel Langley, as well as to the directors of every prominent observatory on the Eastern Seaboard. No astronomer in the country could refuse an invitation to the home of Henry Draper.

It was her home, in fact—her childhood home, built by her late father, the railroad and real estate magnate Cortlandt Palmer, long before the neighborhood became fashionable. Now she made certain the house suited Henry as perfectly as she did, with its entire third floor converted into his machine workshop, and the loft over the stable repurposed as his chemical laboratory, which he reached via a covered walkway connected to the dwelling.

She had barely heeded the stars before meeting Henry, any more than she regarded grains of sand at the shore. He was the one who pointed out to her their subtle colors and differences in brightness, even as he whispered his dream of abjuring medicine for astronomy. If she feigned interest at first to please him, she had long since found her own passion, and proved a willing partner in observation as in marriage. How many nights had she knelt by his side in the cold and dark, spreading foul-smelling emulsion on the glass photographic plates he used with his handcrafted telescopes?

A glance at Henry's plate confirmed he had not touched the banquet fare. He was fighting a cold, or perhaps it was pneumonia. A few weeks earlier, while he and his old Union Army pals were hunting in the Rocky Mountains, a blizzard had struck and stranded them above the timberline, far from shelter. The chill and exhaustion of that exposure still plagued Henry. He looked terrible, as though suddenly an old man at forty-five. Yet he continued chatting amiably with the company, explaining anew, each time anyone asked, how he had generated steady current for the Edison lamps from his own gas-powered dynamo.

Soon she and Henry would be leaving the city for their private observatory upriver at Hastings-on-Hudson. Now that he had finally resigned his professorship on the faculty at New York University, they could devote themselves to his most important mission. In their fifteen shared years, she had seen his landmark achievements in stellar photography win him all manner of acclaim—his 1874 gold medal from Congress, his election to the National Academy of Sciences, his status as a fellow of the American Association for the Advancement of Science. What would the world say when her Henry resolved the seemingly intractable age-old mystery of the chemical composition of the stars?

After bidding the guests good-night at the close of that glittering evening, Henry Draper took a hot bath, then took to his bed, and stayed there. Five days later he was dead.

IN THE OUTFLOW OF CONDOLENCES following her husband's funeral, Anna Palmer Draper drew some comfort from a correspondence with Professor Edward Pickering of the Harvard College Observatory, one of the guests at the Academy gathering the night of Henry's collapse.

"My dear Mrs. Draper," Pickering wrote on January 13, 1883, "Mr. Clark [of Alvan Clark & Sons, the preeminent telescope makers] tells me that you are preparing to complete the work in which Dr. Draper was engaged, and my interest in this matter must be my excuse for addressing you regarding it. I need not state my satisfaction that you are taking this step, since it must be obvious that in no other way could you erect so lasting a monument to his memory."

This was indeed Mrs. Draper's intention. She and Henry had no children to carry on his legacy, and she had resolved to do so on her own.

"I fully appreciate the difficulty of your task," Pickering continued. "There is no astronomer in this country whose work would be so hard to complete as Dr. Draper's. He had that extraordinary perseverance and skill which enabled him to secure results after trials and failures which would have discouraged anyone else."

Pickering referred specifically to the doctor's most recent photographs of the brightest stars. These hundred-some pictures had been taken through a prism that split starlight into its spectrum of component colors. Although the photographic process reduced the rainbow hues to black and white, the images preserved telltale patterns of lines within each spectrum—lines that hinted at the stars' constituent elements. In after-dinner conversation at the November gala, Pickering had offered to help decipher the spectral patterns by measuring them with specialized equipment at Harvard. The doctor had declined, confident that his new freedom from teaching at NYU would allow him time to build his own measuring apparatus. But now all that had changed, and so Pickering repeated the offer to Mrs. Draper. "I should be greatly pleased if I might do something in memory of a friend whose talents I always admired," he wrote.

"Whatever may be your final arrangements regarding the great work you have undertaken," Pickering said in closing, "pray recollect that if I can in any way advise or aid you, I shall be doing but little to repay Dr. Draper for a friendship which I shall always value, but which can never be replaced."

Mrs. Draper rushed to reply just a few days later, January 17, 1883, on notepaper edged in black.

"My dear Prof. Pickering:

"Thanks so very much for your kind and encouraging letter. The only interest I can now take in life will be in having Henry's work continued, yet I feel so very incompetent for the task that my courage sometimes completely fails me— I understand Henry's plans and his manner of working, perhaps better than anyone else, but I could not get along without an assistant and my main difficulty is to find a person sufficiently acquainted with physics, chemistry, and astronomy to carry on the various researches. I will probably find it necessary to have two assistants, one for the Observatory and one for the laboratory work, for it is not likely that I will find any one person with the varied scientific knowledge that was peculiar to Henry."

She was prepared to pay good salaries in order to draw the most qualified men as assistants. She and her two brothers had inherited their father's vast real estate holdings, and Henry had managed her share of the fortune to excellent effect.

"It is so hard that he should be taken away just as he had arranged all his affairs to have time to do the work he really enjoyed, and in which he could have accomplished so much. I cannot be reconciled to it in any way." Nevertheless she hoped to get the work running as soon as possible under her own direction, and "then, when I can buy the place at Hastings where the Observatory is, to do so."

Henry had built the facility on the grounds of a country retreat owned by his father, Dr. John William Draper. The elder Dr. Draper, the first physician in the family to mix medicine with active research in chemistry and astronomy, had died a widower the previous January. His will bequeathed his entire estate to his beloved spinster sister, Dorothy Catherine Draper, who had founded and run a girls' school in her youth to finance his education. It was not yet clear whether Henry's widow would win control of the Hastings property as she wished, and move Henry's Madison Avenue laboratory there, and endow the site as an institution for original research, to be named the Henry Draper Astronomical and Physical Observatory.

"As long as I could I should keep the direction of the institution myself," she told Pickering. "It seems the only suitable memorial I can erect to Henry, and the only way to perpetuate his name and his work."

At the end she entreated Pickering's counsel. "I am so unusually alone in the world, that without feeling that those friends who were interested in Henry's work would advise me, I could not do anything."

Pickering encouraged her to publish her husband's findings to date, since it might take her a long time to add to them. Once again he extended his offer to examine the glass photographic plates on the measuring machine at Harvard, if she would be so good as to send him some.

Mrs. Draper agreed but thought it best to deliver the plates in person. They were small objects, only about an inch square each.

"I may be obliged to go to Boston in the course of the next ten days to attend to some business matters with one of my brothers," she wrote on January 25. "If so I could take the negatives with me and by going to Cambridge for part of a day, if it was convenient for you, could look over the pictures with you, and see what you think of them."

As arranged, she reached Summerhouse Hill above Harvard Yard on Friday morning, February 9, accompanied by her husband's close friend and colleague George F. Barker of the University of Pennsylvania. Barker, who was preparing a biographical memoir of Henry, had been the Drapers' houseguest at the time of the Academy dinner. Late that night, when Henry was seized with a violent chill while bathing, it was Barker who helped lift him from the tub and carry him to the bedroom. Then he bid the Drapers' neighbor and physician Dr. Metcalfe, another dinner guest, to return to the house immediately. Dr. Metcalfe diagnosed double pleurisy. Although Henry of course received the most tender nursing—and showed some brief promise of improvement—the infection spread to his heart. On Sunday the doctor noted the signs of pericarditis, which precipitated Henry's death at about four o'clock Monday morning, the twentieth of November.

MRS. DRAPER HAD VISITED OBSERVATORIES with her husband in Europe and the States, but she had not set foot inside one in months. At Harvard, the large domed building that housed the several telescopes doubled as the director's residence. Both Professor and Mrs. Pickering ushered her into the pleasant rooms and made her feel welcome.

Mrs. Pickering, née Lizzie Wadsworth Sparks, daughter of former Harvard president Jared Sparks, did not aid her husband in his observations, as Mrs. Draper had done, but acted as the institution's vivacious and charming hostess.

An exaggerated though genuine politeness characterized the directorial style of Edward Charles Pickering. If the observatory's financial straits constrained him to pay his eager young assistants meager wages, nothing prevented his addressing them respectfully as Mr. Wendell or Mr. Cutler. He called the senior astronomers Professor Rogers and Professor Searle, and all but doffed his hat and bowed to the ladies—Miss Saunders, Mrs. Fleming, Miss Farrar, and the rest—who arrived each morning to perform the necessary calculations upon the nighttime observations.

Was it usual, Mrs. Draper wondered, to employ women as computers?

No, Pickering told her, as far as he knew the practice was unique to Harvard, which currently retained six female computers. While it would be unseemly, Pickering conceded, to subject a lady to the fatigue, not to mention the cold in winter, of telescope observing, women with a knack for figures could be accommodated in the computing room, where they did credit to the profession. Selina Bond, for example, was the daughter of the observatory's revered first director, William Cranch Bond, and also the sister of his equally revered successor, George Phillips Bond. She was currently assisting Professor William Rogers in fixing the exact positions (in the celestial equivalents of latitude and longitude) for the several thousand stars in Harvard's zone of the heavens, as part of a worldwide stellar mapping project administered by the Astronomische Gesellschaft in Germany. Professor Rogers spent every clear night at the large transit instrument, noting the times individual stars crossed the spider threads in the eyepiece. Since air—even clear air—bent the paths of light waves, shifting the stars' apparent positions, Miss Bond applied the mathematical formula that corrected Professor Rogers's notations for atmospheric effects. She used additional formulas and tables to account for other influential factors, such as Earth's progress in its annual orbit, the direction of its travel, and the wobble of its axis.

Anna Winlock, like Miss Bond, had grown up at the observatory. She was the eldest child of its inventive third director, Joseph Winlock, Pickering's immediate predecessor. Winlock had died of a sudden illness in June 1875, the week of Anna's graduation from Cambridge High School. She went to work soon afterward as a computer to help support her mother and younger siblings.

Williamina Fleming, in contrast, could claim no familial or collegiate connection to the observatory. She had been hired in 1879, on the residence side, as a second maid. Although she had taught school in her native Scotland, certain circumstances—her marriage to James Orr Fleming, her immigration to America, her husband's abrupt disappearance from her life—forced her to seek employment in a "delicate condition." When Mrs. Pickering recognized the new servant's abilities, Mr. Pickering reassigned her as a part-time copyist and computer in the other wing of the building. No sooner had Mrs. Fleming mastered her tasks in the observatory than the impending birth of her baby sent her home to Dundee. She stayed there more than a year after her confinement, then returned to Harvard in 1881, having left her son, Edward Charles Pickering Fleming, in the care of her mother and grandmother.

NONE OF THE PROJECTS UNDER WAY at the observatory looked familiar to Mrs. Draper. Henry's amateur standing and private means had freed him to follow his own interests at the forefront of stellar photography and spectroscopy, while the professional staff here in Cambridge hewed to more traditional pursuits. They charted the heavens, monitored the orbits of planets and moons, tracked and communicated the courses of comets, and also provided time signals via telegraph to the city of Boston, six railroads, and numerous private enterprises such as the Waltham Watch Company. The work demanded both scrupulous attention to detail and a large capacity for tedium.

When the thirty-year-old Pickering took over as director on February 1, 1877, his primary responsibility had been to raise enough money to keep the observatory solvent. It received no support from the college to pay salaries, purchase supplies, or publish the results of its labors. Aside from interest on its endowment and income from its exact-time service, the observatory depended entirely on private bequests and contributions. A decade had passed since the last solicitation for funds. Pickering soon convinced some seventy astronomy enthusiasts to pledge $50 to $200 per year for five years, and while those subscriptions trickled in, he sold the grass cuttings from the six-acre observatory grounds at a small profit. (They brought in about $30 a year, or enough to cover some 120 hours' worth of computing time.)

Born and bred on Beacon Hill, Pickering navigated easily between the moneyed Boston aristocracy and the academic halls of Harvard University. In his ten years spent teaching physics at the fledgling Massachusetts

Institute of Technology, he had revolutionized instruction by setting up a laboratory where students learned to think for themselves while solving problems through experiments that he designed. Pursuing his own research at the same time, he explored the nature of light. He also built and demonstrated, in 1870, a device that transmitted sound by electricity—a device identical in principle to the one perfected and patented six years later by Alexander Graham Bell. Pickering, however, never thought to patent any of his inventions because he believed scientists should share ideas freely.

At Harvard, Pickering chose a research focus of fundamental importance that had been neglected at most other observatories: photometry, or the measurement of the brightness of individual stars.

Obvious contrasts in brightness challenged astronomers to explain why some stars outshone others. Just as they ranged in color, stars apparently came in a range of sizes, and existed at different distances from Earth. Ancient astronomers had sorted them along a continuum, from the brightest of "first magnitude" down to "sixth magnitude" at the limit of naked-eye perception. In 1610 Galileo's telescope revealed a host of stars never seen before, pushing the lower limit of the brightness scale to tenth magnitude. By the 1880s, large telescopes the likes of Harvard's Great Refractor could detect stars as faint as fourteenth magnitude. In the absence of uniform scales or standards, however, all estimations of magnitude remained the judgment calls of individual astronomers. Brightness, like beauty, was defined in the eye of the beholder.

Pickering sought to set photometry on a sound new basis of precision that could be adopted by anyone. He began by choosing one brightness scale among the several currently in use—that of English astronomer Norman Pogson, who calibrated the ancients' star grades by presuming first-magnitude stars to be precisely a hundredfold brighter than those of sixth. That way, each step in magnitude differed from the next by a factor of 2.512.

Pickering then chose a lone star—Polaris, the so-called polestar or North Star—as the basis for all comparisons. Some of his predecessors in the 1860s had gauged starlight in relation to the flame of a kerosene lamp viewed through a pinhole, which struck Pickering as tantamount to comparing apples with oranges. Polaris, though not the sky's brightest star, was thought to give an unwavering light. It also remained fixed in space above Earth's north pole, at the hub of heavenly rotation, where its appearance was least susceptible to distortion by currents of intervening air.

With Pogson's scale and Polaris as his guides, Pickering devised a series of experimental instruments, or photometers, for measuring brightness. The firm of Alvan Clark & Sons built some dozen of Pickering's designs. The early ones attached to the Great Refractor—the observatory's premier telescope, a gift from the local citizenry in 1847. Ultimately Pickering and the Clarks constructed a superior freestanding model they called the meridian photometer. A dual telescope, it combined two objective lenses mounted side by side in the same long tube. The tube remained stationary, so that no time was lost in repointing it during an observing session. A pair of rotating reflective prisms brought Polaris into view through one lens and a target star through the other. The observer at the eyepiece, usually Pickering, turned a numbered dial controlling other prisms inside the instrument, and thus adjusted the two lights until Polaris and the target appeared equally bright. A second observer, most often Arthur Searle or Oliver Wendell, read the dial setting and recorded it in a notebook. The pair repeated the procedure four times per star, for several hundred stars per night, exchanging places every hour to avoid making errors due to eye fatigue. In the morning they turned over the notebook to Miss Nettie Farrar, one of the computers, for tabulation. Taking Polaris's arbitrarily assigned magnitude of 2.1 as her base, Miss Farrar arrived at relative values for the other stars, averaged and corrected to two decimal places. By these means, it took three years for Pickering and his crew to pin a magnitude on every star visible from the latitude of Cambridge.

The objects of Pickering's photometry studies included some two hundred stars known to vary their light output over time. These variable stars, or "variables," required the closest surveillance. In his 1882 report to Harvard president Charles Eliot, Pickering noted that thousands of observations were needed to establish the light cycle of any given variable. In one instance, "900 measures were made in a single night, extending without intermission from 7 o'clock in the evening until the variable had attained its full brightness, at half past 2 in the morning."

Pickering needed reinforcements to keep watch over the variables. Alas, in 1882, he could not afford to hire even one additional staff member. Rather than dun the observatory's loyal subscribers for more money, he issued a plea for volunteers from the ranks of amateur observers. He believed women could conduct the work as well as men: "Many ladies are interested in astronomy and own telescopes, but with two or three noteworthy exceptions their contributions to the science have been almost nothing. Many of them have the time and inclination for such work, and especially among the graduates of women's colleges are many who have had abundant training to make excellent observers. As the work may be done at home, even from an open window, provided the room has the temperature of the outer air, there seems to be no reason why they should not thus make an advantageous use of their skill."

Pickering felt, furthermore, that participating in astronomical research would improve women's social standing and justify the current proliferation of women's colleges: "The criticism is often made by the opponents of the higher education of women that, while they are capable of following others as far as men can, they originate almost nothing, so that human knowledge is not advanced by their work. This reproach would be well answered could we point to a long series of such observations as are detailed below, made by women observers."

Pickering printed and distributed hundreds of copies of this open invitation, and also convinced the editors of several newspapers to publish it. Two early responses arrived in December 1882 from Eliza Crane and Mary Stockwell at Vassar College in Poughkeepsie, New York, followed by another from Sarah Wentworth of Danvers, Massachusetts. Pickering began assigning particular variables to individuals for observation. Although his volunteers lacked any equipment as sophisticated as the meridian photometer, they could compare their variables with other nearby stars, and estimate the brightness changes over time. "If any of the stars become too faint," he advised them by letter, "please send word, so that observations may be attempted here" with the large telescope.

Some women wrote to request formal instruction in practical or theoretical astronomy, but the observatory provided no such courses, nor could it admit curious spectators, male or female, at night. During the day, the director would be only too pleased to show visitors around the building.

Pickering's daytime duties as director required him to correspond regularly with other astronomers, purchase books and journals for the observatory's library, attend scientific meetings, edit and publish the *Annals of the Astronomical Observatory of Harvard College*, oversee finances, answer inquiries by mail from the general public, host visiting dignitaries, and order supplies large and small, from telescope parts to furnace coal, stationery, pens, ledgers, even "water closet paper." Every bit of observatory business demanded his personal attention or at the very least his signature. Only when a blanket of clouds hid the stars could he find a night's sleep.

MRS. DRAPER'S GLASS PLATES demanded examination by daylight. Although Pickering had heard much about these images, and even discussed them with the doctor the night of the Academy dinner in November, he had not seen them till now. He was accustomed to looking at spectra—the separated rays of starlight—through the telescope, using attachments called spectroscopes that former director Joseph Winlock had purchased in the 1860s, when spectroscopy came into vogue. The live view through the spectroscope turned a star into a pale strip of colored light ranging from reddish at one end through orange, yellow, green, and blue to violet at the other. The spectroscope also made visible many black vertical lines interspersed at intervals along the colored strip. Astronomers believed that the breadth, intensity, and spacing of these spectral lines encoded vital information. Though the code remained unbroken, a few investigators had proposed schemes to classify the stars by type, according to the similarities in their spectral line patterns.

On the Draper plates, each spectrum looked like a gray smudge barely half an inch long, yet some contained as many as twenty-five lines. As Pickering viewed them under a microscope, their detail stupefied him. What skill their capture demonstrated, and what luck! He knew of only one other person in the world—Professor William Huggins of England—who had ever succeeded in capturing a stellar spectrum on a photographic plate. Huggins was also the only man of Pickering's acquaintance, aside from Dr. Draper, to have discovered an able astronomical assistant in his own wife, Margaret Lindsay Huggins.

Mrs. Draper agreed to leave her plates in Pickering's care for a complete analysis, and returned to New York. She promised Mrs. Pickering, who was considered one of Cambridge's most accomplished gardeners, to visit again in spring or summer, in the hope of seeing the observatory grounds in full bloom.

Pickering measured each spectrum with a screw-thread micrometer. By February 18, 1883, he could report to Mrs. Draper that he was finding "much more in the photographs than appears at first sight." The computers had plenty to do in graphing the readings from his every half-turn of the screw, then applying a formula and computations to translate them into wavelengths. It became clear that Dr. Draper had demonstrated the feasibility of studying the stellar spectra by means of photography, instead of by peering through instruments and drawing a record of what the eye saw.

Pickering again pressed Mrs. Draper to publish an illustrated account, not merely to establish priority for her husband, but, more important, to show other astronomers the great promise of his technique.

For help with the preparation of the paper, Mrs. Draper asked a noted authority on the solar spectrum, Charles A. Young of Princeton, to contribute an introduction outlining Henry's methods. Meanwhile she catalogued all seventy-eight plates in the spectra series, relying on Henry's notebooks to specify the date and time of each photograph taken, the star name, the length of every exposure, the telescope used, and the width of the spectroscope slit, plus incidental remarks about observing conditions, such as

"There was blue fog in the sky" or "The night was so windy that the dome was blown around."

Pickering summarized the twenty-one plates he had scrutinized in ten tables with explanations. He reported the distances between spectral lines, stating the methodology and mathematical formulas employed to translate line positions into wavelengths of light. He also commented on the similar work being done by William Huggins in London, and ventured to categorize some of Draper's spectra by Huggins's criteria. When he sent his draft to Mrs. Draper for approval, she balked at the mention of Huggins.

"Dr. Draper did not agree with Dr. Huggins," she wrote Pickering on April 3, 1883, concerning two of the stars in the series. Their nearly identical spectra both showed wide bands, which had made Huggins classify the two stars as a single type, but the Draper photographs revealed that one of these stars also had many fine lines between the bands, which set it apart from the other. "In view of this I should not like to accept Mr. Huggins' classification as the standard when Dr. Draper did not agree with it." Although Pickering had seen the abundance of fine lines she described, he found them too delicate for satisfactory measurement.

"You will not I hope be annoyed at my criticism," Mrs. Draper added, "but I feel in publishing any of Dr. Draper's work that I want his opinions represented as nearly as possible, now that he is not here to explain them himself."

The Drapers had met William and Margaret Huggins while visiting London in June 1879, at the Hugginses' home observatory on Tulse Hill. Mrs. Draper recalled Mrs. Huggins as a petite woman with short, unruly hair that stuck straight out from her head as though galvanized. She was half the age of her husband, but a full participant in his studies, both at the telescope and in the laboratory.

The two couples seemed destined to become either rivals or intimates. William gave Henry the benefit of his lengthier experience by offering helpful advice about spectroscope design. He also recommended a new type of dry, pretreated photographic plate that had lately come on the market. There was no need to paint liquid emulsion on these plates just prior to exposing them, and consequently they allowed for much longer exposure times. Before leaving England, the Drapers purchased a supply of Wratten & Wainwright's London Ordinary Gelatin Dry Plates, which proved a boon indeed. They were particularly sensitive to the ultraviolet wavelengths of light, beyond the range of human vision. Unlike the old wet plates, the dry ones created a permanent record suitable for precision measurement. The dry plates gave the Drapers the wherewithal to photograph the spectra of the stars.

THE PAPER ANNOUNCING the stellar spectra findings, "by the late Henry Draper, M.D., LL.D.," appeared in the *Proceedings of the American Academy of Arts and Sciences* in February 1884. Pickering mailed copies to prominent astronomers everywhere. By return mail dated March 12, he received William Huggins's indignant reaction. Huggins found some of Pickering's measurements "very wild," the letter said with emphasis. "I should be glad if you could see your way to look into this, because it would be better that you should discover the error & publish the correction, than that the matter should be pointed out by others. . . . My wife unites in kind regards to you and Mrs. Pickering."

Pickering was certain he had not erred. And, as Huggins had never explicated his measurement procedures, Pickering stood firmly by his own. As they traded charges, Pickering forwarded Huggins's letters to Mrs. Draper.

Now it was her turn to grow indignant. "I felt very sorry," she wrote Pickering on April 30, 1884, "that you should have been subjected to such an ungentlemanly attack, through your interest in Dr. Draper's work." Before returning the letters to Pickering, she took the liberty of copying one, since "it is worth preserving as a curiosity of epistolatory literature."

During this same time, Pickering was seeking assistants who might help Mrs. Draper advance her husband's work to the next stage. He considered former director Joseph Winlock's son, William Crawford Winlock, currently employed at the U.S. Naval Observatory, to be a very likely prospect, but Mrs. Draper rejected him. To her regret, she could not induce her preferred candidate, Thomas Mendenhall, to leave his professorship at Ohio State University. She channeled some of her frustration into the creation of the Henry Draper gold medal, to be awarded periodically by the National Academy of Sciences for outstanding achievements in astronomical physics. She gave the Academy $6,000 to endow the prize fund, and spent another $1,000 commissioning an artist in Paris to fashion a medal die featuring Henry's likeness.

The spring of 1884 brought Pickering new money worries. The successful five-year subscriptions from generous astronomy enthusiasts had run their course, ending the accustomed annual stipend of $5,000. The director was covering various operating expenses out of his own salary, and even so was forced to let go five assistants. In a touching show of solidarity, observatory colleagues took up a collection to retain one of those who had been dismissed, and furnished "part of the required sum," Pickering told his circle of advisers, "from their own scanty means." He appreciated the "extraordinary efforts on the part of the observers, who have performed without assistance the work in which they were previously aided by recorders. This has required an increase in the time spent in observation, and has rendered the work much more laborious. While this evidence of enthusiasm and devotion to science is most gratifying, it is obvious that it cannot long be continued without injury to health. Indeed, the effects of overfatigue and exposure during the long, cold nights of last winter were manifest in more than one instance."

The motto on the Pickering family coat of arms, *"Nil desperandum,"* plus the lifelong habit of his own thirty-seven years, obliged the director to substitute resourcefulness and resilience for despair. He began formulating a means of combining Mrs. Draper's wishes and wealth with the capabilities and needs of his observatory.

"I am making plans for a somewhat extensive piece of work in stellar photography in which I hope that you may be interested," he informed her in a letter of May 17, 1885.

Pickering intended to redirect most of the observatory's projects along photographic lines. His predecessors the Bonds had recognized the promise of photography, and achieved the first photograph of a star in 1850, but the limitations of the wet plates had impeded further attempts. With the new dry plates, possibilities multiplied. Determinations of stellar brightness and variability would surely prove easier and more accurate on photographs, which could be examined, reexamined, and compared at will. A methodical program for photographing the entire sky would transform the painstaking process of zone mapping. As a bonus, these photographs would reveal untold numbers of unknown faint stars, invisible even through the world's biggest telescopes, because the sensitive plate, unlike the human eye, could gather light and aggregate images over time.

Pickering's younger brother, William, a recent graduate of MIT, was already teaching photographic technique there and testing the limits of the art by trying to photograph objects in motion. The twenty-seven-year-old William had consented to assist Edward in a few photographic experiments with the Harvard telescope. One of their pictures yielded 462 stars in a region where only 55 had been previously documented.

The part of Pickering's plan with the greatest potential interest for Mrs. Draper concerned a new approach to photographing stellar spectra. Rather than focus on one target star at a time, à la Draper or Huggins, Pickering anticipated group portraits of all the brightest stars in a wide field of view. To achieve these, he envisioned a new instrument setup combining telescope and spectroscope with the type of lens used in the studios of portrait photographers.

"I think there will be no difficulty in carrying out this plan without your aid," he assured Mrs. Draper. "On the other hand, if it commends itself to you, I am confident that we could make it conform to such conditions as you might impose."

"Thanks for your kindness," she replied on May 21, 1885, "in remembering my desire to be interested in some work with which Dr. Draper's name could be associated, and his memory kept alive. I will be glad to cooperate, if I can, in what you suggest, for its bearing on stellar spectrum photography appeals to me very strongly." More than two years had passed since

Henry's death. Still unable to make his observatory productive, she saw no harm in lending his name to Harvard.

Pickering proceeded slowly and with caution, apprising her of his progress until he could send her some sample images of stellar spectra taken through his new apparatus. She found them "exceedingly interesting." On January 31, 1886, she said, "I would be willing, if the plan could be carried out satisfactorily, to authorize the expenditure of $200 a month or somewhat more if necessary." Pickering thought more would be needed. They settled terms on Valentine's Day for the Henry Draper Memorial—an ambitious photographic catalogue of stellar spectra, gathered on glass plates. Its goal was the classification of several thousand stars according to their various spectral types, just as Henry had set out to do. All results would be published in the *Annals* of the Harvard College Observatory.

On February 20, 1886, Mrs. Draper sent Pickering a check for $1,000, the first of many installments. Pickering publicized the new undertaking in all the usual places, including *Science, Nature,* and the Boston and New York newspapers.

Later that spring Mrs. Draper decided to increase her already generous gift by donating one of Henry's telescopes. She visited Cambridge in May to make the arrangements. Since the instrument needed a new mounting—something Henry had meant to build himself—she asked George Clark of Alvan Clark & Sons to fabricate the parts, at a cost of $2,000, and to oversee the transfer of the equipment from Hastings to Harvard. Once arrived, it would require its own small building with a dome eighteen feet in diameter, and Mrs. Draper meant to cover that expense as well. Together with the Pickerings, she strolled among the plantings of rare trees and shrubs around the observatory to select a site for the new addition.

CHAPTER TWO

What Miss Maury Saw

THE INFUSION OF FUNDS for the Henry Draper Memorial made the Harvard College Observatory hum with new people and purpose. Construction of the small building to house Dr. Draper's telescope started in June 1886 and continued through the summer while Mrs. Draper toured Europe. In October the instrument was mounted in the new dome. Now there were two telescopes outfitted for nightly rounds of spectral photography—the Draper 11-inch and an 8-inch purchased with a $2,000 grant from the Bache Fund of the National Academy of Sciences. The illustrious Great Refractor, through which the first-ever photograph of a star had been taken in 1850, later proved unsuitable for photography. Its 15-inch lens had been fashioned for visual observing; that is, for human eyes most attuned to yellow and green wavelengths of light. The lenses of the two new instruments, in contrast, favored the bluer wavelengths to which photographic plates were sensitive. The 8-inch Bache telescope also boasted a wide field of view for taking in huge tracts of sky all at once, rather than homing in on single objects.

In less than a decade at the helm, Edward Pickering had shifted the observatory's institutional emphasis from the old astronomy, centered on star positions, to novel investigations into the stars' physical nature. While half the computing staff continued to calculate the locations and orbital dynamics of heavenly bodies, a few of the women were learning to read the glass plates produced on-site, honing their skills in pattern recognition in addition to arithmetic. A new kind of star catalogue would soon emerge from these activities.

The earliest known star counter, Hipparchus of Nicaea, catalogued a thousand stars in the second century BC, and later astronomers enumerated the content of the heavens to ever better effect. The projected Henry Draper Catalogue would be the first in history to rely entirely on photographs of the sky and to specify the "spectrum type," as well as the position and brightness, for myriad stars.

Dr. and Mrs. Draper had gathered their spectra one by one, using a prism at the telescope's eyepiece to split the light of each star. Pickering and his assistants, eager to increase the pace of operations, altered the Drapers' approach. By installing prisms at the objective, or light-gathering end of the telescope, instead of at the eyepiece, they were able to capture group portraits containing two or three hundred spectra per plate. The prisms were large, square sheets of thick glass, wedge-shaped in cross section. "The safety and convenience of handling the prisms," Pickering found, "is greatly increased by placing them in square brass boxes, each of which slides into place like a drawer." Harvard's picture gallery grew apace. When Mrs. Draper paid another visit soon after Thanksgiving, Pickering assured her that any star visible from Cambridge appeared on at least one of the glass plates.

Toward the end of December 1886, just when the staff had smoothed out most of the difficulties with the new procedures, Nettie Farrar's beau proposed. Pickering was all in favor of marriage, of course, but he hated to lose Miss Farrar, a five-year veteran of the computing corps whom he had personally trained to measure spectra on the photographic plates. On New Year's Eve, he wrote to inform Mrs. Draper of Miss Farrar's engagement, and also to name Williamina Fleming, the former maid, as her replacement.

Since returning from Scotland in 1881, Mrs. Fleming had been assisting Pickering with photometry. Often she took the director's penciled notations from the nightly observations with his assistants and applied the formulas he specified to compute the stars' magnitudes. By 1886, when the

Royal Astronomical Society awarded Pickering its gold medal for this work, he had already embarked on a parallel approach to photometry via photography. This change required Mrs. Fleming, well accustomed to reading lists of numbers scribbled in the dark, to judge magnitudes from fields of stars on glass plates.

Mrs. Fleming had let Pickering know that photography ran in her pedigree. Her father, Robert Stevens, a carver and gilder praised for his gold-leaf picture frames, had been the first in the city of Dundee to experiment with daguerreotyping, as the process was called in her childhood. She was still a child, only seven, when her father died suddenly of heart failure. Her mother and older siblings tried, for a time, to keep the business running without him, but without success. One by one, her older brothers sailed away to Boston, where she eventually followed them. Now, at twenty-nine, she had a seven-year-old child of her own to care and provide for. Edward would soon arrive; her mother was booking passage with him on the *Prussian* out of Glasgow.

Miss Farrar dutifully introduced Mrs. Fleming to the plates of stellar spectra, and taught her how to measure the hordes of tiny lines. Mrs. Fleming could have taught Miss Farrar a thing or two about marriage and childbirth, but on the subject of the spectrum she had everything to learn.

THE YOUNG ISAAC NEWTON coined the word *spectrum* in 1666, to describe the rainbow colors that arose like ghostly apparitions when daylight passed through cut glass or crystal. Although his contemporaries thought glass corrupted the purity of light by imparting color to it, Newton held that colors belonged to light itself. A prism merely revealed white light's component hues by refracting them at different angles, so that each could be seen individually.

The microscopic dark lines within the stellar spectra, to which Mrs. Fleming now directed her attention, were called Fraunhofer lines, after Joseph von Fraunhofer of Bavaria, their discoverer. A glazier's son, Fraunhofer had apprenticed at a mirror factory and gone on to become a master crafter of telescope lenses. In 1816, in order to measure the exact degree of refraction in different glass recipes and lens configurations, he built a device that combined a prism with a surveyor's small telescope. When he directed a beam of light from the prism through a slit and into the instrument's magnified field of view, he beheld a long, narrow rainbow marked by many dark lines. Repeated trials convinced him that the lines, like the rainbow colors, were not artifacts of passage through glass, but inherent in sunlight. Fraunhofer's lens-testing apparatus was the world's first spectroscope.

Charting his finds, Fraunhofer labeled the most prominent lines with letters of the alphabet: *A* for the wide black one at the rainbow's extreme red end, *D* for a dark double stripe in the orange-yellow range, and so on through the blue and violet to a pair named *H*, and ending farther along the violet with *I*.

Fraunhofer's lines retained their original alphabetical designations through the decades following his death, gaining greater importance as later scientists observed, mapped, interpreted, measured, and depicted them with fine-nib pens. In 1859 chemist Robert Bunsen and physicist Gustav Kirchhoff, working together in Heidelberg, translated the Fraunhofer lines of the Sun's spectrum into evidence for the presence of specific earthly substances. They heated numerous purified elements to incandescence in the laboratory, and showed that each one's flame produced its own characteristic spectral signature. Sodium, for example, emitted a close-set pair of bright orange-yellow streaks. These correlated in wavelength with the dark doublet of lines that Fraunhofer had labeled *D*. It was as though the laboratory sample of burning sodium had colored in those particular dark gaps in the Sun's rainbow. From a series of such congruities, Kirchhoff concluded the Sun must be a fireball of multiple burning elements, shrouded in a gaseous atmosphere. As light radiated through the Sun's outer layers, the bright emission lines from the solar conflagration were absorbed in the cooler surrounding atmosphere, leaving dark telltale gaps in the solar spectrum.

Astronomers, many of whom had considered the Sun a temperate, potentially habitable world, were shocked to learn of its inferno-like heat.

However, they were soon placated—even soothed—by the revelatory power of spectroscopy to expose the chemical content of the firmament. "Spectrum analysis," Henry Draper told the Young Men's Christian Association of New York in 1866, "has made the chemist's arms millions of miles long."

Throughout the 1860s, pioneering spectroscopists such as William Huggins discerned Fraunhofer lines in the spectra of other stars. In 1872 Henry Draper began photographing them. While the number of spectral lines in starlight paled in comparison to the rich tapestry of the Sun's spectrum, several recognizable patterns emerged. It seemed that the stars, which had for so long been loosely categorized by brightness or color, could now be further sorted according to spectral features hinting at their true nature.

In 1866 Father Angelo Secchi of the Vatican Observatory divided four hundred stellar spectra into four distinct types, which he designated by Roman numerals. Secchi's Class I contained brilliant blue-white stars such as Sirius and Vega, whose spectra shared four strong lines indicating the presence of hydrogen. Class II included the Sun and yellowish stars like it, with spectra full of many fine lines identifying iron, calcium, and other elements. Classes III and IV both consisted of red stars, differentiated by the patterns in their dark spectral bands.

Pickering challenged Mrs. Fleming to improve on this elementary class system. Whereas Secchi had sketched his spectra from direct observations of a few hundred stars, she would enjoy the advantage of the Henry Draper Memorial photographs, boasting thousands of spectra for her scrutiny. The glass plates preserved more faithful portrayals of the positions of Fraunhofer's lines than drawings could ever provide. Also, the plates picked up lines at the far violet end of the spectrum, at wavelengths the eye could not see.

MRS. FLEMING REMOVED EACH GLASS PLATE from its kraft paper sleeve without getting a single fingerprint on either of the eight-by-ten-inch surfaces. The trick was to hold the fragile packet by its side edges between her palms, set the bottom—open—end of the envelope on the lip of the specially designed stand, and then ease the paper up and off without letting go of the plate, as though undressing a baby. Making sure the emulsion faced away from her, she released her grip and let the glass settle into place. The wooden stand held the plate in a picture frame, tilted at a forty-five-degree angle. A mirror affixed to the flat base caught daylight from the computing room's big windows and directed illumination up through the glass. Mrs. Fleming leaned in with her loupe for a privileged view of the stellar universe. She had often heard the director say, "A magnifying glass will show more in the photograph than a powerful telescope will show in the sky."

Hundreds of spectra hung suspended on the plate. All were small—little more than one centimeter for the brighter stars, only half a centimeter for the fainter ones. Each had to be tagged with a new Henry Draper catalogue number, and also identified by its coordinates, which Mrs. Fleming determined using the millimeter and centimeter rules inscribed on the wooden plate frame. She read off these numbers to a colleague who sat beside her, penciling the information into a logbook. Later they would match the Henry Draper numbers to the stars' existing names or numbers, if any, handed down from previous catalogues.

In the rune-like lines of the spectra, Mrs. Fleming read enough variety to quadruple the number of star categories recognized by Father Secchi. She replaced his Roman numerals, which quickly grew cumbersome, with Fraunhofer-style alphabetical order. The majority of stars fell into her A category because they displayed only the broad, dark lines due to hydrogen. The B spectra sported a few other dark lines in addition to those of hydrogen, and by her G category the presence of many more lines had become the norm. Type O bore only bright lines, and Q served her as a catchall category for peculiar spectra she could not otherwise pigeonhole.

Pickering applauded Mrs. Fleming's efforts, even as he conceded the arbitrary, empirical nature of her classification. He predicted that in time, with ever more stars studied, the underlying reasons for the different spectral appearances would reveal themselves. Possibly different stellar temperatures were responsible, or different chemical blends, different stages of stellar development, or some combination of such factors—or something as yet unimagined.

In January 1887 Pickering hit on a way to enlarge some of the spectra from smudge-like traces to an impressive four inches by twenty-four. He astonished Mrs. Draper by sending her several examples. "It scarcely seems possible that stellar spectra can be taken which will bear the enlarging of those that you have sent me," she wrote on January 23. "I wonder what Mr. Huggins will say when he sees them." This question stimulated her to strengthen her support of the Henry Draper Memorial, which currently amounted to about $200 a month, by promising $8,000 or $9,000 per year in perpetuity.

There seemed no reason for Mrs. Draper to cling any longer to the dream of continuing her husband's research herself. She thought it best to divest the Hastings observatory of his remaining telescopes, and donate the lot to Harvard. The largest, with its 28-inch-diameter mirror, would likely prove a significant aid in Pickering's pursuits. Still she wavered. It had been one thing to part with the 11-inch-aperture refractor, now ensconced at Cambridge, but the 28-inch reflector preserved precious memories of her wedding day.

Henry had always preferred reflecting telescopes, which gathered light by means of a mirror in lieu of a lens, over refracting ones that could introduce spurious color effects. He had begun crafting his own mirrors right after medical school, and must have made a hundred in all, but the 28-inch was his great reflector. On November 12, 1867, the day after he and Anna exchanged marriage vows in her father's living room, they went downtown together to shop for a glass disk—the kind used in skylights—large enough to form a mirror 28 inches across. They referred ever after to that excursion as "our wedding trip." It took them years to grind and polish the disk to the desired curvature and apply the ultrathin coat of silver that transformed the glass into a perfect mirror.

The 28-inch reflector had enabled them to take their landmark first picture of the spectrum of Vega in 1872, as well as their unrivaled photograph of the so-called Great Nebula in Orion ten years later, and also their final series of stellar spectra images during the summer before Henry's death. On one of those humid July nights, undone by overcast skies, the two of them had quit the observatory around midnight to retire. But as they neared their country house two miles away on Wickers Creek in Dobbs Ferry, they saw the clouds dissipating, so they turned the horses around and drove back to Hastings to resume their work. She remembered returning that way on numerous other occasions just to seize a few more hours—even long ago, when they thought they had all the time in the world.

"MRS. DRAPER HAS DECIDED to send to Cambridge a 28-inch reflector and its mounting," Pickering announced on March 1, 1887, in the first annual report of the Henry Draper Memorial. He praised the project's benefactress for providing not only the instruments required for the project but also the means for keeping them actively employed by operators "during the whole of every clear night," and for "reducing the results by a considerable force of computers," and for publishing them as well. He hoped that other donors would follow her example by similarly endowing astronomy departments elsewhere with the means to function to their fullest.

In the spring of 1887, while Mrs. Draper negotiated with the Hudson River Railroad for a car to carry the 28-inch to Harvard, the observatory received another huge bounty—approximately $20,000, to be augmented by $11,000 annually—for the establishment of an auxiliary station on a mountaintop.

Pickering had been climbing mountains all his life. He began summiting in New England with youthful companions who called him "Pick" and even "Picky." He later measured the heights of points of interest in New Hampshire's White Mountains on solo treks with fifteen pounds of apparatus strapped to his back. In 1876, around the time he left the MIT physics department to direct Harvard's observatory, he founded the Appalachian Mountain Club for fellow outdoorsmen, and served as its first president. Still an active member in 1887, he could well imagine the advantage of stationing a telescope at high altitude.

The source of the sudden windfall was the contested will of Uriah

Boyden, an eccentric inventor and engineer who had received an honorary Harvard degree in 1853. When Boyden died in 1879, unmarried and childless, he allotted $230,000 to perch an observatory far above the atmospheric disturbances that plagued astronomers at sea level. Many noble institutions, including the National Academy of Sciences, vied for control of the Boyden estate, but Pickering convinced Boyden's trustees that Harvard University was the most likely of the suitors to invest the money wisely, and the Harvard Observatory most fit to carry out the testator's instructions. Triumphant after five years of polite wrangling, Pickering organized an exploratory expedition to the Colorado Rocky Mountains.

The Boyden Fund gave Pickering the means to hire his younger brother away from MIT. William, likewise a charter member of the Appalachian Mountain Club, thus became the director's assistant and guide for the western reconnaissance. The brothers left Cambridge in June 1887 along with Lizzie Pickering, three junior volunteers from the observatory, and fourteen crates of equipment. Mrs. Draper joined them at Colorado Springs in July.

Although no high-altitude astronomical observatory yet existed in the United States, the federal reservation at Pikes Peak was home to the world's highest meteorology station, maintained at 14,000 feet by the U.S. Army Signal Corps. This made Pikes Peak the only American mountain where particulars of weather (beyond the statistic of annual rainfall) were known. When Pickering's party of five men ascended in August, leading mules laden with scientific instruments, they encountered a snow squall, a hailstorm, and a thunderstorm they described as violent. Over the course of the month, they camped and compared conditions on three peaks in the region by various means, such as a sunshine recorder William had modified as a complement to a rain gauge, and also by photographing the sky through a 12-inch telescope. Conditions did not seem optimal. What was worse, rumor had it that Pikes Peak might be turned into a state tourist attraction, and be overrun with non-astronomers.

Pickering returned to Cambridge without having settled the placement of the Boyden Station. He thought he might revisit the Rockies the following summer, or try a different mountain range.

In October, after Mrs. Draper returned East, closed her Dobbs Ferry house for the season, and reestablished herself on Madison Avenue, she thanked Pickering for the summer's adventure with the gift of an ornamental pocket telescope that had once belonged to King Ludwig of Bavaria.

WITH TWO AND OFTEN THREE TELESCOPES taking pictures through the night, the observatory devoured plates at a rapid rate. Between 1886 and 1887, advances in the quality of manufactured dry plates extended their recording range to fainter stellar magnitudes, and Pickering took full advantage of each new development. He tried different companies' wares and shifted suppliers accordingly; he encouraged manufacturers to keep improving the sensitivity of their plates—and to send him their latest products for testing.

The volume of data to be calculated rose in proportion to the number of photographs taken. Anna Winlock's younger sister, Louisa, assumed her place in the computing room in 1886, and was joined the following year by Misses Annie Masters, Jennie Rugg, Nellie Storin, and Louisa Wells. The staff of female computers now numbered fourteen, including Mrs. Fleming, who served as their supervisor. Most of the ladies were younger than she, more or less her social equals, and respectful of her authority. That situation shifted in 1888 with the addition of twenty-two-year-old Antonia Maury, who was not only a Vassar College graduate with honors in physics, astronomy, and philosophy, but also the niece of Henry Draper.

"The girl has unusual ability in a scientific direction," Mrs. Draper told Pickering on March 11, 1888, "and is anxious to teach chemistry or physics—and is studying with that object in view."

As a child, Antonia Maury was allowed into her Uncle Henry's chemistry laboratory at the big house in New York City, where she "assisted" him by handing him specific test tubes he requested for his experiments. Before she turned ten, her father, the Reverend Doctor Mytton Maury, an itinerant Episcopal minister, taught her to read Virgil in the original Latin. Her mother, Henry Draper's sister Virginia, was a naturalist enamored of every bird, flower, shrub, and tree on the Hastings property; she had died in 1885 while Antonia was studying at Vassar.

Pickering felt uncomfortable offering the standard computer pay of twenty-five cents per hour to a person of Miss Maury's achievements. He expressed something like relief when she failed to answer his letter, but Mrs. Draper interceded for her through April and May.

"The girl has been very busy," the aunt explained. Although Reverend Maury had relocated to Waltham, Massachusetts, for his work, he had neither found a home for his family nor enrolled his two younger children, Draper and Carlotta, in school, leaving Antonia to take charge of these matters. By mid-June she had joined the Harvard corps.

Pickering assigned Miss Maury the spectral measurement of the brightest stars. Mrs. Fleming had worked from plates containing hundreds of spectra crowded together, and on which the bright stars appeared overexposed. The 11-inch Draper telescope focused on just one star at a time. Each spectrum imaged in this manner spread over an expanse of at least four inches, even before enlarging. The gratifying increase in detail gave Miss Maury much to ponder as she examined the plates under a microscope. In the same blue-violet region of Vega's spectrum where her uncle had photographed four lines in 1879—and ten in 1882—she now counted more than one hundred.

Along with measuring the distances between the lines and converting them to wavelengths, she was expected to classify each spectrum according to Mrs. Fleming's criteria. But Miss Maury had so much more detail to work with that she could not confine her impressions to those parameters. Some of the lines she looked at were not simply thick or intense, but also hazy or fluted or otherwise noteworthy. Such nuances surely deserved attention, for they might illustrate as yet unsuspected conditions in the stars.

WHEN HARVARD'S SECOND MOUNTAIN reconnaissance headed West in November 1888, Pickering opted out. He could not possibly afford enough time away from the observatory to fulfill the mission's ambitious itinerary, which was to begin site testing near Pasadena, California, and continue among the Andes in Chile and Peru. He put his brother, William, in charge. While in California, the team would also visit the Sacramento Valley to observe and photograph the total solar eclipse of January 1, 1889.

Ordinarily, Pickering did not support eclipse expeditions, on practical grounds. He deemed the expense too high, given the high risk of failure. An ill-placed cloud during the scant moments of totality could scotch the whole enterprise (as he had learned firsthand when he went to Spain with former director Winlock for the eclipse of December 22, 1870). But if, as in the present case, the path of totality nearly crossed the path of exploration for the new Boyden Station, Pickering would not object to a small detour.

Favorable weather smiled on the observers for the New Year's Day eclipse. Excitement at the rare sight, however, shook the astronomers and the large crowd of onlookers alike. At the start of totality, the spectators started to yell. The noise drowned out William's call to the person counting out the seconds, and his struggle to make himself heard caused him to take fewer pictures than he intended. He also forgot to remove the lens cap from the spectroscope.

From his disappointment in Sacramento, William went south to Mount Wilson, where he and a few assistants were to test atmospheric conditions by observing for several months with a 13-inch telescope they brought along for that purpose. At the same time, the other half of the team departed for South America. In Pickering's grand scheme, two mountain observatories were better than one. A California aerie would improve on the work done at Cambridge, while an additional satellite station in the Southern Hemisphere would widen Harvard's field of view to encompass the entire sky.

Pickering entrusted control of the South America venture to Solon I. Bailey, age thirty-four, who had joined the observatory staff as an unpaid assistant two years earlier and quickly proven himself deserving of a salary. Like Pickering, Bailey had a younger brother with a talent for photography, and so, with Pickering's blessing, Solon appointed Marshall Bailey as his second-in-command, and planned to meet him in Panama after the eclipse.

Facing a trip expected to last two full years, Solon took along his wife, Ruth, and their three-year-old son, Irving.

The February 1889 voyage aboard the *San Jose* of the Pacific Mail gave Bailey occasion to practice his Spanish with several fellow passengers, whose names he recorded in his journal. On deck, he enjoyed watching Venus sink into the sea after sunset, "plainly seen till she touched the water." In the predawn February sky, he sighted the Southern Cross for the first time. Bailey had loved the stars since his boyhood in New Hampshire, where he witnessed the great natural fireworks of the 1866 Leonid meteor shower. Now he would meet a sky's worth of new constellations, which prospect inured him to whatever hardships lay ahead.

The bulk of the Andes expedition supplies—everything from photographic plates to prefabricated buildings—traveled with Marshall from New York to the Isthmus of Panama, then overland, past the recently aborted French canal effort and the graveyards of fever victims to another ship bound for Callao, near Lima.

The party rode the Oroya Railroad twenty miles east from Lima to Chosica, and from there the Bailey brothers ascended on foot and by mule to elevations of 10,000 feet or more. Their native guides nursed them through bouts of altitude sickness with an effective local remedy, namely the odor of bruised garlic. No particular peak impressed Bailey as ideal, but he needed to seize the good weather of the dry season, and so settled for a nameless mountain with the least obstructed view. It stood just over 6,500 feet high, barely accessible by a path that switchbacked up and around for eight miles. The Baileys labored alongside a dozen locals for three weeks to improve the route from the hotel in Chosica to the site, and then helped drag eighty loads of equipment up that road to the makeshift observatory. When the family moved in on May 8 along with their Peruvian assistant, two servants, cats, dogs, goats, and poultry, their only neighbors were centipedes, fleas, scorpions, and the occasional condor. They relied on a muleteer for daily supplies of water and food.

The Baileys assessed the brightness of the southern stars with the same meridian photometer that Pickering had used in Cambridge, in order to make their observations exactly comparable to his. Similarly, they photographed the southern stellar spectra for the Henry Draper Memorial with the selfsame 8-inch-aperture Bache telescope that had seen nightly duty through the project's first two years. Mrs. Draper replaced the original workhorse at Harvard with another of the same specifications.

Solon Bailey stayed in touch with Pickering as regularly as the mails allowed. When he shipped the first two cases of glass plates to Cambridge, he said they came from an as yet unnamed place that he would like to call Mount Pickering.

"Mt. Pickering might wait," the director wrote back on August 4, 1889, "until I have done as good work as you have on a Peruvian mountain." With local approval, the Baileys christened the site Mount Harvard instead.

When the October onset of the rainy season halted work on Mount Harvard, Bailey moved his wife and son to Lima, then set off with his brother to scout better locations for a permanent base. It took them four months to find a place that met their requirements, on the high desert plain near the town of Arequipa. At 8,000 feet, the air was clear, dry, and steady, and the nearby volcano, El Misti, was nearly extinct.

WHILE THE BAILEYS EXPLORED PERU, Edward Pickering became engrossed with the odd spectrum of a star called Mizar in the handle of the Big Dipper. The star had first drawn his surprised attention on a Draper Memorial photograph taken March 29, 1887, which showed an unprecedented doubling of the spectrum's K line. (Although Fraunhofer's original lettering ended at *I*, later researchers added other labels.) Soon after Pickering shared the unusual news with Mrs. Draper, the strange effect vanished as suddenly as it had appeared. Subsequent images of Mizar's spectrum failed to recover the double K line, but still Pickering kept watching for its return. On January 7, 1889, Miss Maury saw it, too. Pickering, who rarely invoked an exclamation point, wrote Mrs. Draper, "Now it seems nearly certain that it is sometimes double and sometimes single!" Although, he quickly added, "It is hard to say what this means." He suspected that Mizar, also known as Zeta Ursae Majoris, might turn out to be two stars with virtually identical spectra, too closely aligned to be seen separately, even through a big telescope.

Miss Maury could picture the Mizar pair as two wary combatants, circling each other while vying for advantage. Her distant vantage point made it difficult to distinguish the two separate bodies—impossible, in fact, when either one stood in front of the other along her line of sight. But Mizar's twin fighters were emitting light. As they revolved, their relative motions slightly altered the light's frequency: the approaching starlight shifted slightly toward the blue end of the spectrum, the receding starlight toward the red. Those shifts added up to the small K-line separation that created the doubling effect.

Pickering and Miss Maury tracked Mizar's K line through months of ambiguous changes, until they saw the doubled line again on May 17, 1889. Photographs taken a few nights before and after the doubling portrayed the line as hazy—somewhere between single and double. Miss Maury had been wise to trust her intuition about hazy lines.

That Sunday, on her day off, Miss Maury wrote to her aunt, Ann Ludlow Draper, the wife of Henry's brother Daniel. Everything she reported in her long, chatty letter seemed to touch on the theme of single and double. On a visit to the Boston Public Garden she had seen "a wonderful display of tulips single and double of all colors." She now had dual Vassar Alumnae Association membership in both the Boston and New York branches. "I told them I should have a chance to vote twice but they didn't seem to be afraid." She saved the most interesting case for last:

"Tell Uncle Dan that the other day Prof. Pickering succeeded in photographing the double K line of Zeta Ursae Majoris. Other lines were also double that at times are single so I suppose his theory is proved that the change is due to the rotation of two close stars of the same type around one another. It is a very pretty thing. They have been trying for months to catch it double. Prof. Pickering thinks its period must be about fifty days but has not finished the calculations yet. Of course nothing ought to be said about it publicly till it is all worked out." She signed the letter "With love, Antonia."

Pickering wrote a report of the preliminary results, making sure to credit "Miss A. C. Maury, a niece of Dr. Draper" for her careful study of Mizar's spectrum. He sent the paper to Mrs. Draper, who carried it to Philadelphia for the annual meeting of the National Academy of Sciences, where their mutual friend George Barker read it aloud to the assembly on November 13, 1889. Barker assured Pickering that the K-line news "awakened a lively interest."

A few weeks later, on December 8, with Mrs. Draper present at the observatory, Mizar's K line doubled again, right on schedule. Within days, Miss Maury found the double K line in another star, Beta Aurigae (the second brightest in the constellation of the Charioteer). Now there were two examples of newfound star pairs that had been discovered by their spectral characteristics alone. And before the week was out Mrs. Fleming identified a third suspected "spectroscopic binary" on several plates from Peru.

"Now if all these results ensue in consequence of your recent visit here," Pickering cajoled Mrs. Draper, "is it not a sufficient argument in favor of your coming oftener?"

Mrs. Draper wished she might flatter herself, she replied, "that the interesting results obtained during my visit were in consequence of my being with you; my friends have often called me a 'Mascotte' but I fear my luck will not extend so far." Nevertheless she declared herself "delighted" with the new finds. Additional examples would help convince certain members of the Academy, present at the recent meeting, who "thought our imagination had run away with us." More confirmation came in an independent discovery of another spectroscopic binary, also in late 1889, by Hermann Carl Vogel of the Potsdam Observatory.

Vogel had been using spectroscopy to answer a different question—not What are stars made of? or How can stars be divided into groups? but How fast do they move toward or away from Earth in the line of sight? By the degree to which certain lines in their spectra shifted toward blue or red, Vogel calculated their radial velocity. Some traveled as fast as thirty miles per second, or well over one hundred thousand miles an hour.

As Miss Maury continued to chart the spectral changes of Mizar, she concluded that its component stars orbited their common center of gravity once every fifty-two days. She deduced an even shorter period of only four days for Beta Aurigae, the spectroscopic binary that she had discovered. Indeed, she could watch the Beta Aurigae spectrum change from one photograph to the next over the course of a single night. She calculated the orbital speeds in the two binary systems. "A mile a minute" sounded rapid to her ear, but these stars were racing around at more than a hundred miles a second. Her uncle Henry had looked to the spectra to uncover the stars' chemistry, and now the spectra were also yielding the stars' celerity.

THE YEAR 1890 SAW THE PUBLICATION of Mrs. Fleming's opus, "The Draper Catalogue of Stellar Spectra," in volume 27 of the observatory's *Annals.* Pickering rewarded her with a raise in salary and full acknowledgment in his introductory remarks: "The reduction of the plates was begun by Miss N. A. Farrar, but the greater portion of this work, the measurement and classification of all the spectra, and the preparation of the Catalogue for publication, has been in charge of Mrs. M. Fleming." She styled herself "Mina Fleming" now. In addition to the dedication she had shown in measuring and classifying the spectra of ten thousand stars, she had also expertly proofread the catalogue's four hundred pages. Most of the pages consisted of tables, twenty columns wide and fifty lines long, representing approximately one million digits in all.

The Draper Catalogue sorted the stars by the appearance of their spectral lines—not merely for the sake of sorting, but in the hope of opening new avenues of investigation. The classification inspired Pickering, for one, to analyze the distribution of stars by spectral type. Peering into the luminous band of the Milky Way, he found a preponderance of B stars. The B stars clustered along the Milky Way as though they had an affinity for one another or for that region of space. The Sun, a G star, seemed to Pickering to have little relation to the lights of the Milky Way.

Meanwhile Miss Maury proceeded with her own elaborate classification system. She intended to increase Mrs. Fleming's fifteen classes to twenty-two, and also subdivide each type into three or four subcategories, based on the further gradations she detected in the spectra of her bright stars. The strain on her vision prompted her to consult a Boston oculist, who prescribed eyeglasses.

"Dear Auntie," she wrote to her great-aunt Dorothy Catherine Draper on February 18, 1890, "I am now writing up the results of my work of the last two years. I have made a short outline that is the beginning of my classification. I was very much afraid Prof. Pickering would not like it, but I am glad to find that he is quite satisfied and says with a few changes it will do to print. Of course it will take me a long time to get the whole thing written and I expect all the details will make quite a volume. . . . I wear your black hat every day and your afghan keeps me warm at night."

In his fourth annual report of the Henry Draper Memorial, published shortly after Mrs. Fleming's catalogue in 1890, Pickering announced that the total number of photographs taken with the various telescopes had reached 7,883. Other observatories, he noted, made the "very common mistake" of accumulating photographs without deriving results from them through discussion and measurement. At Harvard, however, a corps of computers had been studying the photographs for several years, so that "for many purposes the photographs take the place of the stars themselves, and discoveries are verified and errors corrected by daylight with a magnifying-glass instead of at night with a telescope." Here, too, as in the *Annals,* he cited both Mrs. Fleming and Miss Maury by name. It was the niece of Henry Draper, he emphasized, who had discovered the doubling of the lines in Beta Aurigae.

In line with his usual practice, Pickering distributed the fourth annual report of the Henry Draper Memorial far and wide, including publication in *Nature* and other scientific journals. The report found one of its most appreciative audiences in England, at the home of astronomer and military engineer Colonel John Herschel. As a grandson of William Herschel (discoverer of the planet Uranus) and a son of Sir John Herschel (thrice president of the Royal Astronomical Society), the colonel had seen his share of important leaps in celestial knowledge.

"I have just rec'd your last H. D. Mem. report," he wrote to Pickering on May 28, 1890. "It is very like a pudding all plums—but I will ask you to convey to Miss Maury my congratulations on having connected her name with one of the most notable advances in physical astronomy ever made."

Like the colonel's much celebrated great-aunt, Caroline Herschel, Miss Maury had entered a field of discovery dominated by men, yet she stood among the first astronomers to detect an entirely new group of objects through the upstart method of spectral photography. Its future—and hers—seemed full of promise.

CHAPTER THREE

Miss Bruce's Largesse

EVEN BEFORE SOLON BAILEY selected the site for Harvard's Southern Hemisphere observatory, Edward Pickering had envisioned a superb new telescope to mount there. This ideal instrument would have a lens 24 inches in diameter, or triple the size of the trusty 8-inch Bache, and would therefore gather nine times as much light. He estimated the cost of manufacture at $50,000. In November 1888 he issued a general appeal for the needed funds, and, as in a fairy tale, another heiress stepped forward to grant his wish.

Catherine Wolfe Bruce lived in Manhattan, not far from Anna Draper, but the two were unacquainted before their fortunes crossed in the Harvard Observatory. Miss Bruce, more than twenty years older than Mrs. Draper, had no practical experience with telescopes of any kind. She was a painter and a patron of the arts. Although she lacked Mrs. Draper's knowledge of astronomy, she had long nurtured a vague, distant interest in the subject. Now, at seventy-three, she evinced a genuine eagerness to support further research in the field. As the eldest surviving child of the successful typefounder and print innovator George Bruce, she controlled the disbursement of his wealth. In 1888 she paid $50,000 to erect the George Bruce Free Library on Forty-second Street and fill it with books. An equal expenditure on a single scientific instrument did not seem unreasonable to her, especially the way she heard Pickering describe it when he called on her at home on the morning of June 3, 1889. The large photographic telescope of his dreams, he informed her, would be the most powerful ever pointed at the sky. Dispatched to some lofty mountain for unimpeded, unceasing work, it promised to enrich humankind's knowledge of the distribution and constitution of the stars, far beyond the combined capabilities of numerous—even much larger—telescopes of more typical design.

Perhaps Pickering's reference to the 24-inch object glass as a "portrait" lens appealed to Miss Bruce's artistic sensibility. Surely his optimistic enthusiasm provided an antidote to the disquieting article she had recently read by astronomer Simon Newcomb, director of the U.S. Nautical Almanac Office and professor at the Johns Hopkins University. Professor Newcomb predicted that no exciting astronomical finds would turn up in the near or even the distant future. Since "one comet is so much like another," he asserted "that the work which really occupies the attention of the astronomer is less the discovery of new things than the elaboration of those already known, and the entire systematization of our knowledge."

Miss Bruce viewed the matter differently. Nowhere had she seen a complete list of the ingredients of stars, nor did anyone seem to know what made them shine, or how they formed in the first place. The more she read, the more questions occurred to her. What occupied the spaces between the stars? How could Professor Newcomb call the knowledge complete? As she judged astronomy's prospects, the introduction of photography and spectroscopy, along with advances in chemistry and electricity, suggested that major new findings were afoot. She was counting on Professor Pickering to prove her right, and within weeks of his visit she sent him the requisite sum of $50,000.

As Pickering expressed his thanks to Miss Bruce, he assured his other benefactress that her project, the Henry Draper Memorial, would reap great rewards from the acquisition of the Bruce telescope—at no added cost to the Draper fund.

Mrs. Draper's beloved 28-inch telescope, like the 11-inch before it, had been installed in its own new domed building at the observatory. Although it was the largest of the four telescopes she donated, and the one she had been the most reluctant to part with, it was not living up to expectations. Willard Gerrish, the observatory's talented and innovative tinkerer, along with George Clark, the telescope maker, had spent the first few months of 1889 fussing with it, trying various configurations and adjustments, but wrested from it only a single good spectrum of a faint star. These frustrating experiences increased Pickering's admiration for Dr. Draper's skill, but also forced him to admit defeat, and he abandoned further experiments with the instrument. Mrs. Draper, disappointed but understanding, joined the Pickerings that summer for a short vacation in Maine.

Miss Bruce made no plans to visit Cambridge, as she rarely left home. ("Rheumatism and Neuralgia have racked me badly," she explained.) Nevertheless she followed every step of the telescope's progress via close correspondence with Pickering, beginning in mid-1889, when he ordered the four large lens disks from the firm of Edouard Mantois in Paris. Miss Bruce had learned about glass in her salad days, while collecting art and antiquities on travels throughout Europe. Immersed now in her astronomy self-education, she found the lens for the new telescope preoccupied her as no figurine or chandelier ever had.

"I bought [Charles] Young's Elements of Astronomy," she told Pickering, "after reading in a newspaper that it was adapted to the humblest capacity— Well there is in 'every lowest depth a lower deep' and I fear to fall into it.

"Young calls the vast spaces between the stars a vacuum," Miss Bruce continued, while another book she read by philosopher John Fiske "speaks of it as the luminiferous ether. I shall hold on to Young." Pickering obligingly provided her with all the Harvard Observatory's publications, from volumes of the *Annals* to offprints of his research reports. "Your paper on Variable Stars of Long Periods," she said in a thank-you note, "I at once read and with admiration— not of the Tables but of the simple goodness of heart shown in the detailed directions to unskilled amateurs how to become useful aids to Science."

Since his initial 1882 open invitation to amateurs, especially ladies, to observe the changing brightness of variable stars, Pickering had repeated the request with relevant instructions, and also rewarded the volunteers by publishing several summaries of their results in the *Proceedings of the American Academy of Arts and Sciences.* He recommended that amateurs follow only those variables that cycled slowly through their brightness changes over periods of days or weeks, and leave the more rapid or erratic examples to study by professionals. No amount of amateur assistance, however, relieved Pickering of the need to repeat his exhortations for additional funding in every annual report of observatory activities.

Upon hearing that certain millionaires had failed to open their pocketbooks in response to a worthy appeal, Miss Bruce reminded Pickering that "some generalship is required" in dealing with rich gentlemen: "They must not be attacked directly and squarely but in flank or rear." For her part, she volunteered to lend further assistance, not just to Harvard, but to astronomers everywhere, if Pickering would agree to help her choose the most deserving cases. With her promise of $6,000 to start, he announced a call for aid applications in July 1890. He also sent letters to individual researchers at observatories all over the world, asking whether they could put $500 to immediate good use—say, to hire an assistant, repair an instrument, or publish a backlog of data. Nearly one hundred responses met the October deadline. Pickering evaluated the proposals and Miss Bruce approved his recommendations in time for a November selection of the winners. Simon Newcomb, author of the article that had aroused Miss Bruce's indignation, became one of the first five scientists in the United States to receive her support. Another ten awards went overseas to astronomers working in England, Norway, Russia, India, and Africa.

"The same sky overarches us all," Pickering avowed when he submitted the list of awardees to the *Scientific American Supplement.* As usual, he hoped that word of one donor's generosity would spur others to follow suit. But no one proved more motivated by the outcome than Miss Bruce herself. She felt a particular obligation to astronomers whose plans had arrived too late for consideration.

"My dear Professor," she wrote Pickering on February 10, 1891, "I am sorry that so lately as the date of your letter, Jan. 10th, applications still came in, and to see clearly that mixed with some good we have done some harm, for these are disappointed persons, even in some cases mortified—though in fact without cause." She urged Pickering to assess a new crop of astronomers whose projects she could assist.

All this time, Miss Bruce's lavish gift to Harvard still lay in the bank unused, awaiting the arrival of the lens disks from Paris. Pickering's queries to the glassmaker, Mantois, went unanswered, as did letters and cablegrams sent from the Clarks. After eighteen months, Miss Bruce denounced "that miserable laggard Mantois," and wished she could confront him in person, confident that her command of French was "probably at least as good as his."

In the spring of 1891, nearly two years after Pickering placed the lens order, he discovered to his distress that Mantois had not even begun to form the glass.

"I shall be only less glad than you when the disc arrives and Clark finds it satisfactory," Miss Bruce sympathized on April 9. "Let your patience hold out a little longer—another two years or so—and what are two years in the calculations of an astronomer?"

WILLIAM H. PICKERING, the designated first director of Harvard's southern observatory, reached Arequipa in January 1891. He viewed his arrival as the foundation of a dynasty. His brother already ruled the familiar realm of the northern skies from Cambridge, while here below the equator William would explore the lesser known heavens and establish his own reputation. True, he supervised only two astronomical assistants for the moment, but he presumed the need for a larger staff in Peru would become apparent as soon as the rainy season ended and observations commenced.

William first had to lease or buy land in the area the Bailey brothers had scouted. Solon and Ruth Bailey were packing to go home, vacating their rented house in Arequipa so the Pickerings could move into it. William had come accompanied by his wife, Anne; their two toddlers, Willie and Esther; Anne's widowed mother, Eliza Butts of Rhode Island; plus a nurse. To accommodate his family in accord with his sense of mission, he treated the $500 sum he had been allotted for land acquisition as merely the down payment on an expensive property. There he began construction of several permanent buildings for the telescopes, and also a commodious hacienda, complete with servants' quarters and stable. In February, after only a few weeks in residence, William cabled Edward, "Send four thousand more."

By Western Union and stern letters in longhand, Edward tried to make William hew to a stricter economy. In addition, the older brother repeatedly pressed the younger to get busy taking pictures. The Henry Draper Memorial hungered for more photographs of southern stellar spectra. Why did William not make use of the Bache telescope already set up on-site, even as he oversaw the erection of shelters for the three additional telescopes he had brought to Peru? (Over a comparable period during the first expedition in 1889, Bailey had returned some four hundred plates.) In April, William finally obeyed, but still delayed sending the photographs to Cambridge. By August, Edward complained in exasperation, "I am very glad that you have 500 plates but very sorry that they are not here. I am very anxious lest some mistake regarding instructions may make them worthless."

William had never been happier, never enjoyed better seeing—the astronomer's term for atmospheric conditions. He loved the clear, still mountain air of the Andes that enabled him to resolve unprecedented fine detail on the surfaces of the Moon and planets. Although the solar system was not the focus of any Harvard program planned for Peru, the planets now absorbed William's attention almost to the exclusion of photometry and spectroscopy. Despite his early devotion to photographic technique, William backslid into visual observing at Arequipa. The 13-inch Boyden telescope, with which he photographed the eclipse in California, had suffered some damage to its clock drive on the journey south, rendering it temporarily unfit for long-exposure photography. Until new parts were in place, William felt free to savor the view through the instrument. It had a reversible lens that rendered it equally fit for the eye or the camera. Even after the needed repairs to the 13-inch were completed, and it stood ready to photograph the spectra of the brightest southern stars, William preferred to peer through its eyepiece and sketch the landscape of Mars.

While William neglected his duty in Peru, Mantois in Paris honored other lens orders ahead of Harvard's. Miss Bruce deputized J. Cleaves Dodge, an old family friend living in France, to visit the glazier in the hope of rousing him to action on her telescope.

"We are not in luck," Miss Bruce told Pickering on October 1, 1891, "decidedly not— Accept my condolences. Here is another cause of delay— Before you see all those discs you will have discovered your first grey hair and I! I shall be in cool repose in Greenwood [Cemetery]. But read Mr. Dodge's letter."

The enclosure described a cordial, half-hour conversation in which M. Mantois explained to Mr. Dodge "the mysteries of Crown and Flint glass, which to manufacture and to manipulate, as he seems to do, one must be a real alchemist." This was hardly an exaggeration. Telescope lenses required glass made from the highest-quality materials, mixed according to secret recipes, and heated for weeks at temperatures above one thousand degrees in guarded foundries. The terms "crown" and "flint" distinguished the two basic types of glass by the added quantities of lead in the latter. Used alone, either crown glass or flint glass yielded lenses that brought different wavelengths of light to different focal points, creating a jumble of color distortion known as chromatic aberration. United, however, crown and flint corrected each other. As Joseph von Fraunhofer demonstrated in the early nineteenth century, a "doublet," formed by a convex lens of crown glass paired with a concave complement of flint glass, could bring the focal points into better alignment.

"The trouble in the making of the lenses," Dodge's report to Miss Bruce continued, "seems to be the numerous accidents that occur in the firing and baking of the very best specimens, and which no human intelligence can foretell." Mantois had lost months to bad luck with a 40-inch lens commissioned by another university and could not yet say for certain when he might satisfy Harvard, willing though he was. Dodge reproduced a verbatim recital of the man's plight: "M. Mantois said, 'You see I am as interested as anyone in the completion of the work, for I am not paid anything till it is all finished, but I can only send that which is perfectly satisfactory. Besides

I am constantly in a great state of anxiety as to the baking of the molds; I have tubes connected with my bed to warn me at night if the fires are cooling; and the falling asleep of one of the watchmen may cost me no end of trouble and expense.'" Dodge left Mantois's establishment convinced that no other career in manufacturing "is attended with more chances of failure than this one of glazier for telescopes."

HAVING CLASSIFIED TEN THOUSAND STARS, Mina Fleming turned her organizational gift to the arrangement of the ever-multiplying glass plates. The myriad photographs filled many wooden chests, shelves, and cupboards in both the computing rooms and the library. She imagined they would soon exceed all available space in the observatory building. In the interim she filed them by telescope and by type—the chart plates that mapped each section of the sky, the group spectra, the individual bright spectra, the star trails, and so on—each one in a brown paper envelope, each envelope labeled by number, date, and other identifying details, all of which were repeated on index cards in a card catalogue. Rather than pile the plates in columns, she stood them on edge for easy access. Reason to revisit one or another stored plate arose daily as the assistants examined, measured, discussed, and performed computations upon each new batch of photographs. When, for example, Mrs. Fleming spotted a spectrum that struck her as characteristic of a variable star, she did not need to wait for future observations to confirm her hypothesis. The evidence of the past would bear her out in the now. She had only to consult her records to see which photographs included that portion of the heavens, then pull the relevant plates from the stacks and compare the star's current state with all its previous manifestations.

"So you have, ready to hand and for your immediate use," Mrs. Fleming pointed out in a summary of her method, "the material for which a visual observer might have to wait" a very long time, perhaps indefinitely. Moreover, the plates trumped any visual observer's report, "for in the case of the observer, you have simply his statement of how the object appeared at a given time as seen by him alone, while here you have a photograph in which every star speaks for itself, and which can at any time, now or in the years to come, be compared with any other photographs of the same part of the sky."

Early in 1891, after she had identified a new variable in the constellation of the Dolphin, and, with the director's approval, published her finding in the *Sidereal Messenger,* two skilled observers from other institutions took it upon themselves to corroborate the discovery. Both contested her claim, declaring the star *not* variable. When those same two astronomers met to discuss their conclusions, however, they realized they had each been watching a different star, neither of which was in fact Mrs. Fleming's star. "No such error," she all but crowed, "could have occurred from the comparison of the photographic charts."

Detecting new variable stars had become Mrs. Fleming's forte. Although fewer than two hundred such inconstant lights were known when she joined the observatory staff, the decade of her employment flushed out a hundred more, of which she personally identified a score. She made her earliest finds while gauging magnitudes by the size of the speck a star created on a photographic plate, and then noting which specks changed size in subsequent pictures. Spectra gave her an easier means. Once she had familiarized herself with the spectral features of a few well-known variables, she could recognize similar traits in other stars, almost at a glance. For example, the presence of a few light hydrogen lines among the black ones signaled a variable star near the height of its brightness.

As Mrs. Fleming ferreted out new variables, she also kept a close watch on the old. The director was keen to monitor how the spectra of variable stars changed over time, and the ways that variations in brightness correlated with the appearance of the Fraunhofer lines.

In the spring of 1891, Mrs. Fleming noticed something unusual about the familiar variable called Beta Lyrae. Its changeable nature had been known for a hundred years, but now, looking at its magnified spectrum, she recognized the doubled lines signifying that Beta Lyrae belonged to the newly defined group of spectroscopic binaries—that this star was in fact two stars.

Miss Maury also took an interest in Beta Lyrae, even a proprietary interest, given that Lyra (the Harp) was a northern constellation, and she had charge of the approximately seven hundred brightest stars of the northern skies. Together with Pickering and Mrs. Fleming, she reviewed twenty-nine Draper Memorial plates that contained images of Beta Lyrae. Her analysis suggested this binary did not comprise identical twins, as was the case for Mizar and Beta Aurigae, but two stars of different classes, each varying at its own rate and for its own reasons. She began to frame a theory about the nature of their relationship.

Pickering had hoped to publish Miss Maury's classification of the northern bright stars by the end of 1891, as a sequel to Mrs. Fleming's 1890 "Draper Catalogue of Stellar Spectra." Unfortunately, Miss Maury seemed nowhere near ready to release her results. Her two-tiered classification system, which addressed both the identity and the quality of the spectral lines, required a painstaking exactitude. Anything less would deny the complexity of the problem. Although her slow pace disturbed Pickering, he could hardly accuse her of slacking. She had taken on a second job as a teacher in the nearby Gilman School, while still pursuing her observatory work so assiduously that he feared she neglected her health. Mrs. Draper, too, grew impatient with her niece. After a visit to the observatory in early December, she wrote Pickering, "I do hope Antonia Maury will make an effort and finish more satisfactorily what she has in hand."

Pickering stopped daily by the computing room to monitor the assistants' progress. Miss Maury shrank from these encounters. She often went home feeling tired and nervous, and more than once complained to her family that the director's criticism had shaken her faith in her own ability. Incapable of continuing under such conditions, she quit the observatory early in 1892. Through the next few months she negotiated with Pickering about the fate of her unfinished projects, which she refused to abandon or cede to anyone else.

"I have had in mind for some time to explain to you," she wrote on May 7, "how I feel in regard to the closing up of my work at the Observatory. I am willing and anxious to leave it in satisfactory condition, both for my own credit and in honor of my uncle. I do not think it is fair to myself that

I should pass the work into other hands until it is in such shape that it can stand as work done by me. I do not mean that I need necessarily complete all the details of the classification, but that I should make a full statement of all the important results of the investigation. I worked out the theory at the cost of much thought and elaborate comparison and I think that I should have full credit for my theory of the relations of the star spectra, and also for my theories in regard to Beta Lyrae. Would it not be fair that I should, at whatever time the results are published, receive credit for whatever I leave in writing in regard to these matters?"

Pickering stood ever ready to credit her. He just wished he had some idea of when that occasion might arise.

MISS MAURY'S DEPARTURE at the start of 1892 coincided with the long-awaited arrival from France of the Bruce telescope's glass disks, two of flint and two of crown, each two feet in diameter by three inches thick, weighing in the neighborhood of ninety pounds, and rimmed in a metal hoop. The flawless purity of the glass rendered the disks invisible, and therein lay their beauty. Pickering immediately consigned them to the Clarks for the all-important grinding and polishing. He expected the transformation of the disks into the four-element portrait lens to take at least six months of long days on the Clarks' steam-powered lathe. First the glasses would be abraded with rough sand, then by ever-finer rouge powders, until they assumed the desired curvature.

While that process was under way, Pickering drew plans for a freestanding structure in which to assemble and try out the finished instrument. The Bruce telescope must pass his own stringent tests before he could ship it to Arequipa. And Arequipa, in turn, must be readied to receive it. On May 29 he notified William, who had disappointed him, that his term as southern director would expire at the end of the year, at which time Solon Bailey would replace him. William could return in future to observe at the site, if he liked, but he would no longer be in charge.

William recoiled at the insult. "Without being boastful, I think I've accomplished a pretty big thing," he argued on June 27, 1892, "and if the authorities [the president and fellows of the Harvard Corporation] could see it they would say I had got them a great deal for their money." The idea of subservience to Bailey particularly rankled William: "As to our coming down here again to Peru and living in a small hut, while the Baileys occupy the Director's house, it is out of the question. I planned and built that house, and while I am in Peru I expect to live in it. I don't choose to live in a shanty while one of my subordinates occupies the house I built."

All through the summer of 1892, William soothed himself by studying Mars during its close approach. As he reported in *Astronomy and Astro-Physics,* he observed and drew the red planet every night save one from July 9 to September 24. He collected "considerable data" on the Martian polar caps, the shaded areas "of greenish tint," and the two large, dark regions that, under favorable conditions, turned blue "presumably due to water." He referred to these as "seas." He corroborated the numerous Martian "canals" originally discovered by Giovanni Schiaparelli of Italy, and noted that many of them intersected one another—at junctions he dubbed "lakes." William communicated these same findings to the editors of the *New York Herald,* who printed them to sensational effect. An exasperated Edward Pickering complained to William on August 24 that the waters of Mars had generated a "flood" of forty-nine newspaper cuttings in one morning. He admonished William to restrict himself "more distinctly to the facts."

Meanwhile Edward and Lizzie Pickering were looking to remodel the "dwelling house" in the observatory's east wing. Although they had no children, nor any personal need for extra space, they expanded the observatory apartments, at their own expense, to accommodate and entertain visiting astronomers. Pickering was content to have the college continue docking his $4,000 annual salary for amounts considered rent, but he asked that henceforth the monthly sums be allocated solely for the observatory's use, instead of for Harvard at large, as had been customary. Despite frequent gifts from active donors and the receipt of important new bequests, the director feared it might take years for the budget to recover from William's profligacy in Peru.

Miss Bruce, unaware of William's indiscretions, followed his publications in the astronomy literature. "The two articles in the May number of AstroPhysics from the pen of your brother," she wrote Pickering in August, "have given me great pleasure and caused me to reflect on the happiness that you must have in working thus into each other's hands." She imagined Edward and William to be as close to each other as she was with her sister Matilda, ten years younger, who lived with her and helped her in a hundred ways.

The following month gave both Pickering and Miss Bruce genuine cause for shared happiness. "I hold out my hand to grasp yours," she effused on September 9, when she heard that the lenses for the large photographic telescope had passed their first examination. "Let us rejoice."

In October, as though in atonement, William resumed photography at Arequipa for the Henry Draper Memorial. By the end of December 1892 he had shipped two thousand plates to Cambridge.

ALMOST FROM THE MOMENT stars began amassing on Harvard's glass photographic plates, the director developed a dread of their destruction by fire. The larger the collection grew, the more devastating the contemplation of its loss, should the wooden observatory building ignite. Virtually everyone of Pickering's acquaintance had lost something of value to a conflagration. Mrs. Draper's family, for one, owned a theater in Union Square that burned to the ground in 1888, and its reconstruction continued to cause her grief. Consequently she had become something of an expert on fireproof paint, periodically urging its application to the observatory.

Pickering favored an alternate solution. In 1893 he announced the completion of a two-story "fire-proof building," made entirely of brick, for the safe storage of glass plates and manuscripts of yet-to-be published results. The Brick Building, as everyone soon came to call it, crowned Pickering's fifteen years of site improvements, from the numerous telescope domes and sheds to the neighboring house on Madison Street that had been transformed into a photography workshop and darkroom. In the words of journalist Daniel Baker, whom Miss Bruce commissioned to write up the observatory's history, the hilltop once dominated by a single edifice had become a "little city of science."

Mrs. Fleming oversaw the packing of the thirty thousand plates into three hundred crates. On March 2, 1893, workers rigged a block and tackle from the roof of the observatory's west wing to a window of the new repository. Then they slid the approximately eight tons of plates down the rope skyway at the rapid clip of a crate per minute. Despite the precarious flight, not one piece of glass cracked or shattered.

Naturally Mrs. Fleming and most of the computers followed the plates into the new space, to remain close to them. They traveled at ground level by a wooden walkway over the muddy intervening ditch. When Miss Maury returned to join them in the spring, Pickering asked for her promise to complete her classification before the end of the year or turn over the work to someone else, and she signed a statement saying that she would.

There were now seventeen women computing at the observatory. In other words, nearly half of the observatory's forty assistants were female—a fact Mrs. Fleming intended to emphasize in her invited remarks for the upcoming Congress of Astronomy and Astro-Physics in Chicago.

The name of the congress called attention to astronomy's increasing emphasis on the physical nature of the stars through spectroscopy. Some self-styled astro-physicists were already distancing themselves from the more traditional observers who concentrated on stellar positions or cometary orbits. George Ellery Hale trumpeted the new trend. He had been briefly associated with Harvard while a student at MIT, before establishing his own Kenwood Observatory in his native Chicago in 1890. It was Hale who prevailed upon the editor of the *Sidereal Messenger* to change the publication's name to *Astronomy and Astro-Physics* in 1892. And it was again Hale who organized the August 1893 Congress of Astronomy and Astro-Physics. By timing the meeting to coincide with the Chicago World's Fair, or Columbian Exposition, he added incentive for astronomers from either coast and other continents to undertake the journey.

Hale invited Pickering to present the opening address to fellow scientists at the conference, as well as a broader, less technical talk to inform the fair-going public about the fabric of the stars. Hale also requested an exhibit's worth of photographs documenting the work of the Harvard College Observatory and its physical plants in Cambridge and Arequipa. Pickering included photographs of the women at work in the new Brick Building.

Pickering began preparing the text for his popular address well in advance. "Our only knowledge of the constitution of the stars," it began, "is derived from a study of their spectra."

Mrs. Fleming also prepared an invited paper for the Astronomy and Astro-Physics congress. The previous summer in Chicago had seen the two women's rights federations merged into one "National American Woman Suffrage Association." This year, soon after the Exposition opened in May 1893, suffragettes Julia Ward Howe and Susan B. Anthony had made impassioned presentations. Though Mrs. Fleming fully affirmed the principle of equality, she was not an American citizen, and the feminist struggle for the right to vote was not her fight. The cause she championed was equality for women in astronomy: "While we cannot maintain that in everything woman is man's equal," Mrs. Fleming averred in her Chicago contribution, "yet in many things her patience, perseverance and method make her his superior. Therefore, let us hope that in astronomy, which now affords a large field for woman's work and skill, she may, as has been the case in several other sciences, at least prove herself his equal."

The White City of the Columbian Exposition, with its two hundred grand structures, held numerous fascinations for Anna Draper, who visited the fair in mid-June. The Woman's Building had been designed by Sophia Hayden, the first of her sex to receive a degree in architecture from MIT, and its interior bore murals and paintings executed by well-known female artists such as Mary Cassatt. Other not-to-be-missed highlights included the Electricity Building's seventy-foot-tall tower of lightbulbs and the Hall of Agriculture's fifteen-hundred-pound copy—in chocolate—of the Venus de Milo. Inside the Manufacturers' Building, Mrs. Draper stared up at the mammoth mounting pier and tube of a new telescope that would soon move to a permanent home on the shores of Lake Geneva in Wisconsin.

The tube stood empty. Its 40-inch object glass—the very monster that had vied with the Bruce lens for priority in Mantois's Paris establishment—still lay hundreds of miles back East, on the lathe at Alvan Clark & Sons.

By late summer, progress on the Bruce telescope had reached a critical stage. Only William Pickering was free to represent the Harvard Observatory at the astronomy conference in Chicago. When Mrs. Fleming's speech was read aloud for her at the session held Friday, August 25, William seconded her statements in praise of the efficient women's force in Cambridge. The next day he presented his own report, titled "Is the Moon a Dead Planet?," in which he answered his own question with an emphatic "No."

In early September the first piece of giant iron superstructure for the Bruce telescope made its slow way up Summerhouse Hill. Placement of the two-ton bed plate occupied six men and four horses for a full day. Edward Pickering watched the "ponderous affair" of assembly wear on for two more months before he got the proof he needed to declare the whole grand giant-telescope enterprise entirely worthwhile.

"We have obtained some remarkable photographs," he wrote Miss Bruce on November 19. "I can now safely report its assured success, and can congratulate you on having the finest photographic telescope in the world."

CHAPTER FOUR

Stella Nova

NOTHING IN THE SKY SURPRISED an astronomer more than the sudden apparition of a new star where none had been seen before. When the legendary Tycho Brahe of Denmark glanced skyward one night and beheld such a sight, he declared it "the greatest wonder that has ever shown itself in the whole of nature since the beginning of the world." *De nova stella,* Tycho's eyewitness account of the 1572 marvel, argued that Aristotle had been wrong to call the heavens immutable. Surely the abrupt appearance of the new star and its subsequent disappearance a year later proved that change could occur in the realm beyond the Moon.

Not long after Tycho died in 1601, another nova burst into splendor. Both Galileo in Padua and Johannes Kepler in Prague observed the brilliant new star of 1604, which was so bright as to be visible in the daytime for more than three weeks. Although no comparable naked-eye nova ever materialized over the following centuries, a few fortunate astronomers who happened to be pointing their telescopes to the right place at the right time discovered seven more novae between 1670 and 1892. Then Mina Fleming found one. On October 26, 1893, while hunched over her light lectern with a magnifying glass during a routine perusal of a photographic plate newly arrived from Peru, she seized on a star with the peculiar spectrum unique to a nova—a dozen prominent hydrogen lines, all of them bright.

The director cabled the exciting news to Solon Bailey, who had taken the photograph more than three months earlier, on July 10. Pickering hoped new pictures by Bailey would disclose what, if anything, remained of the nova. Meanwhile Mrs. Fleming looked back in time through the plates to see what had preceded it, but found no trace in prior photographs of the same region. The star must have been dim indeed before its leap from obscurity to seventh magnitude.

The nova lay in a constellation defined and named in the mid-eighteenth century by Nicolas Louis de Lacaille, a French astronomer, on a voyage south. Where others might have seen beasts or deities, Lacaille perceived instruments of modern science, from Microscopium and Telescopium to Antlia (air pump) and Norma (originally Norma et Regula, for the surveyor's square and rule). Now, thanks to Mrs. Fleming, the small, inconspicuous Norma gained fame as the home of the first nova to be detected by spectral photography. It was only the tenth such star to have been observed in recorded history, and it was hers.

Nova Normae's most recent predecessor, the new star of 1891, had been espied visually through the telescope of an Edinburgh amateur, who alerted the Scottish astronomer royal by anonymous postcard. The timely aviso allowed observatories in Oxford and Potsdam to photograph the nova within days of its discovery. Now Pickering placed a picture of that nova's spectrum next to Nova Normae's. The two were virtually identical. Together, they made the ideal illustration for the announcement of the new discovery "by Mrs. M. Fleming," which Pickering submitted in early November to *Astronomy and Astro-Physics.* "The similarity of these two new stars is interesting," he pointed out in his article, "because if confirmed by other new stars it will indicate that they belong to a distinct class resembling each other in composition or physical condition." Even more important, their similarity had enabled Mrs. Fleming to make the discovery, and might lead her to others as she continued sifting through the spectra collected for the Henry Draper Memorial.

Pickering regarded the nova—any nova—as the ultimate variable star. Novae figured first among the five types of variables he defined. Just as astronomers had divided the multitudes of stars into color or brightness or spectral categories in the ongoing effort to comprehend their nature, so, too, the rarer variable stars could be grouped by their behavior. A nova, a "new" or "temporary star," flared and faded just once in a lifetime. Its brief glory thus distinguished Type I from the "long-period" variables of Type II, which underwent the slow, cyclical changes of one or two years' duration, monitored by Pickering's volunteer amateur corps. Type III experienced only slight changes, not easily followed via small telescopes; Type IV varied continuously in short time spans; and Type V revealed themselves to be "eclipsing binaries," or pairs of stars that periodically blocked each other's light.

One could only wonder at the cause of a nova's rapid rise to brightness. Something—a stellar collision, perhaps?—made the star release and ignite enormous quantities of hydrogen gas. The spectra of the two recent novae presented perfect portraits of incandescent hydrogen. Had Pickering become aware of the outburst sooner, instead of fifteen weeks after the fact, he might have tracked Nova Normae through its slow decline, watching the bright lines fade to dark, and the spectrum resume the semblance of a normal star.

SOLON BAILEY SUFFERED NO REMORSE at not having noticed Nova Normae himself. He had been entrusted with the day-to-day operation of the Arequipa station, the nightly rounds of photography, and the timely transfer of photographic plates to Harvard. Although he looked at every image to make sure it passed muster, the detailed scrutiny fell, as always, to the Cambridge staff of assistants and computers. He gladly added his voice to the chorus of congratulatory wishes now surrounding Mrs. Fleming.

Since Bailey's return to Arequipa in late February 1893, he had grown enamored of the great globular clusters of stars visible in the pristine southern skies. These objects, each a mere fuzz patch or hazy star to the unaided eye, appeared through a field glass as globes of nebulous light, dense at the core and fading gently toward the borders. Viewed through the 13-inch Boyden telescope, such clusters resolved into swarms of stellar bees. The abundance of individual components challenged Bailey to take a census of their populations. He began by capturing a single cluster in a two-hour exposure made the night of May 19, 1893. On a separate glass plate he ruled lines to produce a grid of four hundred tiny boxes. Laying the grid over the glass negative, and placing the pair under a microscope, he counted the stars in each compartment. "The cross-hairs of the eyepiece divided each square into four sub-squares," he reported to *Astronomy and Astro-Physics* in June, "which served to prevent confusion in counting." Even so, he asked Ruth Bailey to count, too, for confirmation. When he saw that his wife's tally somewhat exceeded his own, he averaged their results to arrive at a total of at least 6,389 stars in the cluster called Omega Centauri. "There can be no doubt, however," he added, given the difficulty of assessing the closely packed center, "that the whole number of stars comprising this splendid cluster is very much greater." Then he proceeded to gauge the brightness of the individual cluster members, one row at a time, by comparing each star to its neighbors, in sequence—8.7, 9.5, 8.8, 8.5, 9, 8.8, 9.2, and so on.

Bailey thought he might devote his life to the study of clusters, but not at the expense of his regular duties. He kept up the steady flow of chart plates and spectra plates. He outfitted a new meteorology station—the world's highest—at the summit of El Misti with the help of his older brother, Hinman. Their younger brother, Marshall, disaffected by the exhausting work of the initial Peruvian expedition, had declined a second stint at Arequipa and enrolled instead in the College of Physicians and Surgeons in Baltimore.

The globular clusters soon proved themselves fertile hunting grounds for variable stars. Mrs. Fleming picked out the first one in Omega Centauri in August, and Pickering found another a few days later. As these discoveries multiplied, a malcontent from within the Harvard ranks undermined their validity by attacking the observatory's procedures.

Seth Carlo Chandler, a variable star aficionado, had served under Pickering from 1881 to 1886 as a research associate and calculator of comet orbits. After leaving his post, he continued his affiliation with the observatory by helping to issue telegraph alerts of comet sightings and other time-sensitive information to the global astronomy community. In 1888 he released a catalogue of variable stars, complete with his own detailed numerical analyses of their variability. Like Pickering, he appreciated and encouraged the contributions of amateur volunteers to the study of variables, but he differed with the director on the best methods for discovering such stars. Chandler preferred the time-honored techniques of visual observing. Because he distrusted detections made via spectral photography, he omitted nearly all of Mrs. Fleming's recent finds from his second variable star catalogue, published in 1893. Adding further insult in a supplement, he characterized more than a dozen of her discoveries as "alleged but unconfirmed." Worse, in February 1894, in the respected international journal *Astronomische Nachrichten,* Chandler impeached the integrity of the entire Harvard Photometry study published in the observatory's *Annals.* He cited fifteen "serious errors" in the monitoring of variable stars with Pickering's meridian photometer. In each of these cases, the magnitude listed for a given date conflicted with other reliable observers' reports, or with the known pattern of the variable in question, indicating that the photometer had been focused on *the wrong star.* Possibly the instrument was fatally flawed. If it *never* pointed reliably, then *mis*identification might be rampant, and the work worthless.

A colleague of Chandler's digested the charges for public consumption in the pages of the *Boston Evening Transcript* on March 17, 1894, asserting that "adverse statements so sweeping and from so well-known an authority as Dr. Chandler call for an explanation which shall be satisfactory to scientific men."

It was said of Pickering that he loved to discuss but refused to dispute. Forced to make some rejoinder, he wrote a brief letter to the *Transcript*'s editor, printed March 20. He called the attack "unwarranted," adding that the questions raised in it were "scientific in their character" and therefore "unsuited to a discussion in a daily journal." He promised a full reply "through the proper channels." Meanwhile the press in New York and Boston continued to harp on the story.

Mrs. Draper heard of the fracas firsthand from Pickering and also read all about it in the *New York Evening Post.* It struck her as ludicrous for Chandler to assail Pickering's photometric work—work that had been rewarded with the gold medal of the Royal Astronomical Society, the Henry Draper Medal of the National Academy of Sciences, and the Benjamin Valz Prize from the French Academy of Sciences. In her opinion, Pickering's achievements had excited Chandler's jealousy.

The May 1894 issue of the *Nachrichten* carried Pickering's official response. He conceded that the fifteen variable stars pointed out by Chandler had indeed been wrongly identified in the *Annals,* but they were isolated and understandable instances. As for Chandler's broader accusation, well, "It is somewhat as though it should be argued from a physician's losing twenty percent of his cholera patients that he had been equally unfortunate in his general practice."

Newspapers nevertheless kept up their coverage of "Astronomers at War" through the summer months. Harvard president Charles Eliot defended the observatory throughout. On July 31 he cautioned Pickering, "As I have said to you before, the best way of meeting this and all other criticism is to issue more fresh good work, and this I doubt not that you are bent on doing. My chief anxiety in connection with this matter is that it should not disturb your peace of mind or impair your scientific activity. At first it had to a little; but I hope the temporary effect is wearing off. If it does not, I beg to repeat what I said to you at our last conversation—you ought to take a good vacation."

The Pickerings' prescribed vacation in the White Mountains of New Hampshire restored some of the director's equanimity. He felt even better that fall, when a new photometric catalogue from the Potsdam Observatory appeared. It showed near-perfect agreement with the myriad magnitude determinations made at Harvard.

WILLIAM PICKERING, HAVING RELUCTANTLY relinquished his house and position of authority in Arequipa, returned from Peru via Chile, where he observed the total solar eclipse of April 16, 1893. As soon as he resettled in Cambridge, he began plotting his next rendezvous with Mars. Favorable orbital alignments coming up in October 1894 offered William the irresistible opportunity to build on his observations of 1892. It had been his good fortune to find himself ideally situated south of the equator for the last close approach. This time the American Southwest offered the most desirable perspective. Luckily for William, the wherewithal for mounting a trip to the Arizona Territory came to him in the person of Percival Lowell. The wealthy Lowell had recently developed a passion for planetary astronomy, and required an expert's guidance for his first serious endeavor in the field. A Boston Brahmin and Harvard alumnus, Lowell knew the Pickering brothers socially through the Appalachian Mountain Club.

Edward Pickering granted William a year's leave without pay to join Lowell's "Arizona Astronomical Expedition." He also allowed Lowell the yearlong lease of a 12-inch Clark telescope and mount for $175 (a sum equal to 5 percent of the equipment's value). Lowell and William successfully negotiated with another telescope maker, John Brashear of Pittsburgh, for the loan of a second, larger instrument—an 18-inch refractor—to further their cause. On July 14, a euphoric William wrote Edward from Flagstaff to say the seeing in Arizona rivaled that at Arequipa.

At Arequipa itself, Bailey tried to estimate the danger to the Harvard station posed by the opening salvos of civil war in Peru. The country was still rebuilding itself, settling its international debts and internal turmoil after years of fighting as Bolivia's ally in conflicts with Chile. As early as July 1893, Bailey had half-jokingly proposed "to remove the lenses and use the telescope tubes for cannon" if the need arose. Two months later, after taking serious stock of his available defenses ("two or three revolvers"), he concluded that the wisest move in the event of an armed attack would be to surrender "and rely on the government for indemnity." He laid in extra provisions as a precaution and built heavy wooden shutters for the windows and doors. These were not quite complete when rioting and shooting broke out in Arequipa, bringing government troops into the city. After the death of President Francisco Morales Bermúdez in Lima in April 1894, increasing violence prevented the vice president's succession to office. Bailey added an adobe wall between the station and the road, and then another wall along the northern perimeter, facing in the direction of a village that was now rebel-occupied territory. Rebels also controlled the area surrounding the original observing site on Mount Harvard.

Spring elections restored a former president, Andrés Avelino Cáceres, to office in summer, but the political situation remained unstable. The observatory carried on its normal activities to the extent possible. In early September, assistant George Waterbury set out, as he did every ten days or so, to check on the weather gauges installed atop El Misti. When he reached the 19,000-foot summit, he found the meteorology shelter had been vandalized and several of the instruments stolen.

"Dear Uncle Dan," Antonia Maury wrote to Daniel Draper, the Central Park meteorologist, on September 2, 1894, from North Sydney, Nova Scotia, "I have been having a good time here and have got well rested in the last three weeks. I am still however too lazy to be able to make any plans for the winter. I have to be in Cambridge for about two weeks to finish up some odds and ends. Then Mrs. Fleming is going to attend to the printing of the work, so I shall be free. I think a little of going with Carlotta [her sister] to study at Cornell, but may decide to study by myself in Boston where I can have excellent library advantages."

She had missed the agreed-upon deadline of December 1, 1893, for completing her work at the observatory, but felt close to finishing now. Unfortunately, the remaining "odds and ends" overwhelmed her, especially as she also resumed her teaching duties for the semester. Her father, the Reverend Mytton Maury, whose lack of a permanent posting no doubt added to his daughter's stress, expressed his concerns to Pickering on November 12. "I wish you would try to give Miss Maury every assistance in finishing up the work in hand," he wrote. "It is most important that she should go away. She is growing so nervous that she often wakes long before daybreak & can't get to sleep again." Along with the increase in her anxiety from September to

November, her winter plans had taken the shape of a trip to Europe. "She and her brother are to sail on the 5th of Dec.," Reverend Maury said with emphasis. "You will see therefore that a conclusion must be reached. As to the Orion lines please assume that labor yourself & so relieve her. That at least seems to be one point in which her responsibility can be lightened. I do not know that there is anything else that can be done by others—but if there is, please do me the favor to have it done."

The Orion lines, as the reverend must have known from his daughter's description, were particularly conspicuous spectral lines in some stars of the constellation Orion, the Hunter. Orion lines were separate from the twenty known hydrogen lines, distinct also from the calcium lines, and not to be confused with the hundreds of "solar lines" typical of the Sun's spectrum. In short, it was not yet clear what substance or condition the Orion lines represented, but they figured importantly in the first five stellar spectra categories of Miss Maury's classification system.

"It is very desirable to have the work done of course," Reverend Maury continued, "but not at the expense of injured health." In a postscript, he asked Pickering to provide a letter of introduction to foreign astronomers for Miss Maury's use in Europe. Pickering did as he was asked.

"Many thanks for the letter of introduction," Reverend Maury wrote again on December 1. "It was just the thing. . . . Thanks too for your efforts to facilitate the work on those perplexing Orion lines. I hope now things will be left in such a shape that there will be no perturbations in the mind of 'the Astronomer,' as we call her."

Over the next several weeks, as the day of her departure was delayed and Miss Maury continued working at the observatory, she took offense at some remark of the director's, so that Reverend Maury felt it necessary on December 19 to remind Pickering that his daughter "is a lady and has the feelings and rights of one."

In an effort to excuse her father's intervention, Miss Maury sent her own agitated note to Pickering on December 21: "The fact is that my father was excited because I often came home tired and nervous and sometimes complained as people are apt to do about their work. It is true I have often said that your criticisms had from the beginning so shaken my faith in my own ability to work with accuracy that I had been struggling against a great weight of discouragement from the start. But although I several times before have taken offense at things you have said to me I have always decided in the end that the only trouble was that I, being naturally unsystematic, was not able to understand what you wanted and that you also, not having examined minutely with all the details, did not see that the natural relations I was in search of could not easily be arrived at by any cast iron system."

She drafted one last letter while riding the train to New York on January 8. "I am very sorry I did not see you to say goodbye," she began. The last week had passed in such a rush. Her steamer was leaving the next day. "I felt the more sorry as I wanted to tell you that I appreciate your kindness to me all along and understand entirely many things that I did not always [understand] in times past. And that I should have done differently had I seen more clearly. I am sorry I have been so long about the work, but partly on account of my inexperience and partly because the facts developed gradually, I am not sure that I could have done any better what I have done in the past year and six months, at any earlier time." She hoped he would have no trouble reading her manuscript, and promised to send Mrs. Fleming an address in Europe where she could receive mail.

"I sail tomorrow at 2 pm—at least I believe so though I am not sure whether or not I am dreaming, so confused is everything in my mind. I hope that although my work at the observatory is at an end I may still keep your friendly regard and confidence which I value very greatly."

ASTRONOMERS WHO HAD DOUBTED William Pickering's impressions of Mars were scandalized at what Percival Lowell saw there—not just watery surface features, but a fully developed network of irrigation canals engineered by intelligent Martians. William would not go so far. By November 1894 he had made up his mind to leave Lowell and return to the Harvard fold. The choice proved wise, as the weather in Flagstaff that winter destroyed the quality of the seeing.

In Peru, where the seasons were reversed, Solon and Ruth Bailey spent a few overcast January days in 1895 tending to a problem at an auxiliary meteorology station in Mollendo. On their way back to Arequipa, a crowd of armed men surrounded their train and rushed aboard. "The car was at once filled with cries of 'Jesus Maria' and 'Por Dios,' by the ladies and children," Bailey wrote Pickering on January 14. "I advised Mrs. Bailey and Irving to keep quiet and there would be no harm done and so it turned out. The revolutionists behaved with great moderation and offered us no indignity whatever. We were sent back to Mollendo however while the men followed us in another train which they had captured. When near the town they left us locked in the car and forming in line marched in and took the place in a few minutes. Mollendo is said to have a population of about 3000 but there were only 15 soldiers and they surrendered after about a hundred shots were fired."

The Baileys and scores of other temporarily displaced passengers found shelter for the night at the home of the steamship agent. The next day, when the rebels left and troops loyal to President Cáceres reclaimed Mollendo, the Baileys again boarded the train for Arequipa. At home they found that Hinman Bailey had removed the lenses from the several telescopes—not to use the tubes as cannon, as Solon had quipped, but to bury the glass for safekeeping. The Bruce photographic telescope, with its 24-inch lens, was still undergoing tests in Cambridge, and for once the delay in its delivery seemed providential.

Within a fortnight of the train incident, Arequipa came under heavy attack. Rebels cut the telegraph line and Bailey reburied the recently retrieved telescope lenses. In the diary-like letter he composed during the siege, which lasted from January 27 to February 12, he recorded daily events, the din of nearby rifle fire, and his relief that the battle coincided with the cloudy season, "as otherwise it would sadly interfere with our night work."

By March the victorious rebels had ousted Cáceres and installed a provisional government. New elections planned for August seemed likely to elect the rebel leader, Arequipa native Nicolás de Piérola. The Baileys had reported hearing shouts of "Viva Piérola!" punctuating their January ride on the hijacked train. Now they invited the old warhorse to tour the observatory station, and treated his entourage to a reception with refreshments. "The expense was moderate," Bailey assured Pickering on April 15, "about twenty dollars, and as Pierola is sure to be the next president, if he lives, I think it was a wise act."

With good weather and nightly observations restored, Bailey resumed his contemplation of the gorgeous globular clusters. Four of them contained such astonishing numbers of variable stars that he took to calling them "variable star clusters." With Ruth's help, he kept count of their contents as he searched for additional examples.

Pickering promised to send more experienced, more reliable assistants to Peru. Soon he would send the Bruce telescope as well. He had taken more than a thousand photographs with it and worked out the various kinks inherent in its unusual design. For example, the huge tube (truly a piece of heavy artillery) had tended to flex slightly under its own weight, so that long exposures stretched some star images into oblong shapes. The Clarks helped Pickering add strengthening rods and otherwise ready the Bruce to meet its destiny at Arequipa.

The telescopes in Cambridge, in contrast, faced a dim future as the growing city encroached on the observatory. Municipal plans to widen nearby Concord Avenue for streetcars concerned Pickering, for fear the traffic might rattle the Great Refractor atop its several-hundred-ton supporting pier of granite blocks set in gravel and cement. Already the unwanted glare of electric lights thwarted the instrument's power. It could no longer register faint objects such as small comets and nebulae. Pickering had written to various city offices with his concept for screens that could be placed over outdoor light fixtures to prevent them from illuminating the atmosphere above, but the idea fell on deaf ears. Since he could neither eliminate nor shield the streetlights, he learned to make use of their intrusion. "The electric lights," he told the observatory's Visiting Committee of patrons and advisers, "prove an advantage in one way." He and his telescope assistants needed to assess and reassess the clarity of the sky many times per night, so that the quality of the photographs made during each hour could be graded accordingly. Photometry demanded still more rigorous attention to sky conditions, with updates made every few minutes while manning the meridian photometer, when even the faintest wisp of cloud might throw off a brightness reading by several tenths of a magnitude. The streetlights alerted the observers to virtually invisible clouds. "The effect is like that of the Moon," Pickering explained, "but as the lights are below the clouds instead of above them, the latter become conspicuous even when too faint to be seen in moonlight."

THE LETTER OF INTRODUCTION that Pickering had provided for Miss Maury won her a warm welcome at the observatories of Rome and Potsdam. As she traveled abroad with her brother in 1895, Scottish chemist William Ramsay released the results of his laboratory experiments with cleveite gas, which findings threw Miss Maury's Orion lines into stark new relief.

Ramsay, working at University College in London, collected the gas bubbles given off when the uranium compound called cleveite was dissolved in sulfuric acid. He described the properties of the gas and submitted a sample to spectrum analysis. One of its spectral lines shared the same wavelength as a line previously seen only in association with the sun—a line that English astronomer Norman Lockyer attributed in 1868 to a solar substance, which he called helium after the Greek sun god, Helios. Ramsay's new discovery proved that helium occurred on Earth as well. He went on to demonstrate its presence not only in uranium ores but also in the atmosphere.

While Lockyer had named helium on the basis of a single spectral line, Ramsay revealed the element's full spectrum. Its additional lines matched the "Orion lines" that Miss Maury had so often mentioned in the manuscript she left with Pickering upon her departure. She thought it imperative to incorporate the new revelation about helium into her classification, now in preparation for publication. On the other hand, the time for making major revisions had long since passed. "I do not know," she wrote "in haste" in an undated letter to Mrs. Fleming, "whether Professor Pickering will care to insert the statement in regard to Orion lines being due to helium."

SOLON BAILEY TRAVELED ALONE to Cambridge to claim the Bruce telescope in the summer of 1895. Pickering wanted him to spend a few months at Harvard familiarizing himself with the operation of the instrument before superintending its removal to Peru.

Ruth Bailey had asked her husband to carry two gifts to her friend Lizzie Pickering, but the bulky alpaca shawl and robe took up so much room in his luggage that she sent them on ahead, with a letter. "The only regret I have about the robe is that it needed cleaning, and as there are no establishments here for anything of the kind, I was obliged to send it just as it was." She hoped it would reach Cambridge before the Pickerings left for Europe. She also wanted to plead, woman to woman, for Mrs. Pickering to look out for Solon. "I am very anxious for Mr. Bailey to leave Cambridge before December for fear of the cold," she wrote. "I trust you will see that he starts for Arequipa before it is too cold. Men take no care of themselves, that is most men need looking after, they never think they must be careful of their health. I dread to have him go, still I think it is wiser for him to see the instrument there in running order."

Her concerns sounded like typical wifely worries, but the turn of events in the following months lent them eerie prescience. In July, while her husband was at Harvard, their son, Irving, fell seriously ill. Bailey rushed back to Arequipa as soon as he received her cable, though even "as the crow flew," the distance to Peru exceeded four thousand miles, and the roundabout route by available transport widened that gap. Fortunately, the child recovered soon after his father's return.

On February 13, 1896, Bailey stood waiting at the dock to greet the Bruce telescope when its ship pulled into Mollendo. Willard Gerrish had dismantled the instrument in Cambridge and chaperoned the pieces as far as New York, where he took pains to delay loading them until the incoming tide raised the steamer to the level of the wharf. Then he convinced the captain to store the lenses in the vessel's strong room for the long voyage down the eastern coasts of both Americas, through the Strait of Magellan, and up the Pacific to Peru.

Pickering dictated the all-water route, despite its added expense, to avoid the overland shortcut across the Isthmus of Panama. The fewer changes of conveyance through inexperienced hands, the better, he reasoned. Neither Pickering nor Gerrish ever imagined how the steamer would pitch about in Mollendo's harbor, even in the best of weather, or how the waves would toss the little launch that ferried the Bruce piecemeal from ship to shore. The captain laughed as he recounted the extreme care exercised at New York, and Bailey shared the joke with Pickering. "It does look rather risky," he wrote of the Bruce's off-loading, "to see the heavy pieces roll up and down over the heads of the boatmen." The process took a full day but met with no mishap. After reaching Arequipa by train, the telescope ascended the last leg of its journey in an oxcart, along the winding trail to the mountain lookout.

Bailey built a shelter for the Bruce with a dome of canvas-covered wood and a pier of local stone set in mortar as its steady base. By the end of May, after many tests of his perseverance and skill, he achieved images of a quality that pleased him. Just when he thought the instrument's trials were over, the Bruce took an unexpected jolt that almost toppled it.

"Yesterday we had the strongest earthquake which I have ever experienced," Bailey wrote to Pickering on June 15, 1896. "It came at 10:05 A.M. I could distinctly see the ground move, something which I never saw before. I was in the laboratory. I rushed into the Bruce building which was near to see the effect. The whole mass of the castings etc swayed visibly and the tube shook violently." Bailey was pleased to report, however, that all the station's telescopes emerged from the shaking unscathed.

CHAPTER FIVE

Bailey's Pictures from Peru

EDWARD PICKERING HAD COME TO VIEW Solon Bailey as the heir apparent to the Harvard throne. "You are more familiar with the work of the Observatory in general than anyone else," the director assured Bailey soon after he resumed stewardship of the Arequipa station, "and as you have the necessary executive ability I want to make your position one of increasing responsibility." Pickering was not yet fifty, and not looking to retire, but he foresaw the possibility of a sabbatical year for himself, or other protracted absences. He hoped that Bailey, after finishing his current five-year term in Peru, would "undertake more and more of the executive work" at Cambridge, and assume "a large part of the general management of the Institution." But such forecasting was strictly between the two of them, and premature to boot. Pickering could still rely on loyal, amiable Professor Arthur Searle, ten years his senior, to stand in for him whenever necessary.

Searle had first served as acting director after Joseph Winlock's death in 1875, and ran the observatory until Pickering took it over eighteen months later. He had been a classics scholar while a Harvard student, then a Colorado sheep farmer, an English teacher, a clerk in a Boston broker's office, a tutor, and a computer for the U.S. Sanitary Commission. When his astronomer brother, George Mary Searle, left the Harvard Observatory in 1869 to take holy orders as a Catholic priest, Arthur took the vacant place at the telescope. He expected this employment to be as temporary as any of his previous positions, but instead he settled in. A methodical and reliable observer, Searle became adept at photometry, especially as it applied to planetary satellites, asteroids, and comets. He also computed the orbits of these objects and recorded all the observatory's meteorology data. In 1887 he was named Phillips Professor of Astronomy, and taught classes at the nearby Society for the Collegiate Instruction of Women, which became Radcliffe College in 1894.

Although Pickering viewed the observatory as strictly a research facility, he was himself a gifted educator. He had allowed a few determined female students into his MIT physics classes, and instituted the women's astronomy courses early in his tenure at Harvard. By now he felt justly proud of the several alumnae who held "positions of the first importance" in their fields. They included Mary Emma Byrd, director of the Smith College Observatory, and Sarah Frances Whiting, professor of physics and director of the observatory at Wellesley College.

Qualified Radcliffe astronomy students occasionally landed unpaid assistantships at the Harvard Observatory. In 1895 Searle and Pickering selected Henrietta Swan Leavitt for this honor, and Annie Jump Cannon a short while later. These two ladies displayed maturity well beyond that of the typical matriculant. They had both completed college studies, traveled abroad, and done a bit of teaching before assuming their duties on Observatory Hill, where they met each other for the first time. By an odd, unfortunate coincidence, Miss Leavitt was suffering the gradual loss of her hearing during this period, and Miss Cannon, having survived a severe case of scarlet fever while at Wellesley, was already quite deaf.

Pickering put Miss Leavitt to work on a new project in photometry. His own ongoing work in this field entailed nightly observations of stellar brightness with telescopes and photometers, but she was to assess magnitudes from the glass photographs. He gave her several years' worth of plates taken at Cambridge with the 8-inch Bache and 11-inch Draper telescopes, centered on the northernmost stars. Pickering had long relied on the polestar alone as his touchstone, bringing Polaris into the proximity of other stars via mirrors and prisms. Miss Leavitt needed to identify numerous new benchmarks among stars fixed in place on glass plates, and also compare them over time with sixteen long-period variables in the polar region. Later, the visual and the photographic judgments could be cross-checked, calculated, and rectified to achieve a strict new standard of consistency.

Seated at her light lectern, Miss Leavitt chose one variable as a starting point and then proceeded from star to star, judging each one's magnitude, jotting the brightness number right on the glass plate. In the observatory record books she always used pencil, as protocol required, and changed a ledger entry, if necessary, by putting a line through it and setting her corrected value alongside the original, because erasures were forbidden on those pages. But different rules governed the plates. The non-emulsion side of the glass offered a smooth writing surface, where the colors of India ink stood out against the black-on-white images of star fields, and where errors could be wiped away with a handkerchief. When Miss Leavitt reached the end of one stellar pathway, she picked up another, and marked a new trail of stars. Streams of her colorful numbers emanated from the variables like small bursts of fireworks.

It seemed as though each comparison star struck its own particular note in the chorus of light, while some of the variables covered a wide range of several octaves. Miss Leavitt could still think in musical terms, even as the sound of music faded from her perception. She continued every Sunday at church to sing the hymns that had filled her childhood, first in Lancaster, Massachusetts, where she was born on the Fourth of July, 1868; and later in Cleveland, where the family moved when her father, the Reverend Doctor George Roswell Leavitt, became pastor of that city's Plymouth Congregational Church. In Ohio, she spent her seventeenth year enrolled at the Oberlin Conservatory of Music, before the onset of her hearing problems diverted her course. After conservatory, as a student of liberal arts at coeducational Oberlin College, and all through her four years of women's college in Cambridge, she excelled at mathematics, from algebra to geometry to calculus.

Pickering found Miss Leavitt to be of an extremely quiet and retiring nature, absorbed in her work to an unusual degree. Nevertheless he asked her, in February 1896, to introduce the newcomer, Miss Cannon, to the variable stars near the pole. Miss Cannon would also be examining them—not during the daytime on photographs, but nightly by telescope, as the first female assistant ever commissioned to do so. The thirty-two-year-old Miss Cannon owed this privilege to her educational lineage: At Wellesley, she had studied physics with Pickering's MIT protégée Sarah Frances Whiting, in a program of hands-on laboratory instruction modeled on the methods he innovated. Miss Cannon had also taken Professor Whiting's astronomy course, which taught her how to handle Wellesley's 4-inch Browning telescope and kept her abreast of activities at the Harvard College Observatory. When the Great Comet of 1882 arrived like a white-winged bird in the autumn of Miss Cannon's junior year, Miss Whiting supervised observations of its flight over a period of months. For nearly a week the object flared brightly enough to catch the naked eye, even in daylight, but only the telescope could disclose how the comet nucleus broke into pieces after its close brush with the Sun.

Miss Cannon might have followed a quicker trajectory from Wellesley to Harvard, but the lingering effects of scarlet fever grounded her at home in Dover, Delaware. After graduation she took up photography, tutored small groups of students in arithmetic and American history, and made the rafters "ring," as she put it, playing the organ for Sunday school at the Methodist church. A pleasant decade passed in this fashion, until her mother's death plunged her into despair. "Am still here in my little room, surrounded by my memories," she wrote in her diary on March 4, 1894, nearly three months after the funeral of Mary Elizabeth Jump Cannon. "My mother is ever before me. I can see how people lose their minds, for I believe I shall if I am not aroused by something. . . . She was mine and always will be, my most precious mother. Twelve weeks ago tonight, she was downstairs in the spare room and was more worried about my taking a nap on my couch without covering than about herself. She said she knew I was going to be sick for I looked so, and here I am after twelve weeks of agony such as I shall never have to pass through again, here I am well, my constitution will carry me through many weary years, yet may I be led into a useful, busy life. I am not afraid of work. I long for it. What can it be?"

Just as Mrs. Draper had taken solace in founding the memorial project at Harvard after the loss of her husband, so Miss Cannon found a way through her grief by participating in that project. She returned to Wellesley in 1894 as assistant to Miss Whiting, who smoothed her transition to Searle's "Practical Research" class at Radcliffe and a perch in the observatory.

"Soon it will be '97. And three years have passed," Miss Cannon noted at 11:15 p.m. on December 31, 1896, picking up her diary again after a long hiatus. "Two busy years at Wellesley and this one at the Harvard Observatory. The busy life I so longed for has been opened up to me. Friends have come to me from the great world and my heart, my life are now the study of astronomy. They little know what it means to me, how it was the only thread holding my reason, almost my life. . . . I no longer look forward with dread. The days have no terror. I long for my mother just the same, but I feel that I have the patience to run my race, to do the work set before me, and am able to find contentment in my surroundings. I could not help it, thrown as I am with such kind people."

Her colleague Miss Leavitt had taken advantage of a travel opportunity and left the observatory, at least temporarily, but Miss Cannon still counted eighteen fellow female and twenty-one male associates in her new professional family. At night, when weather permitted, she did what had been considered man's work: she used the 6-inch telescope in the observatory's west wing to check on the variables assigned to her, noting the date and hour of each magnitude appraisal she made. Over time, such isolated glimpses would add up to a star's full cycle of variation, or "light curve," from maximum to minimum and back to maximum brightness. The curve, in turn, could suggest the type of variation—and perhaps hint at the cause as well. Whenever Miss Cannon felt unable even to estimate a magnitude, she recorded the reason, such as *c* for cloudy or *m* if bright moonlight impeded her task.

Daylight hours found her at a light lectern alongside the women in the computing room, examining photographic plates from Arequipa. Miss

Cannon's allotted niche in the Henry Draper Memorial concerned the spectra of the brightest southern stars. The director wanted her to create a southern counterpart to Miss Maury's classification of the bright stars of the north. The 13-inch Boyden telescope in Peru provided Miss Cannon with the same kind of widely dispersed, highly detailed spectra that the 11-inch Draper telescope had produced for Miss Maury. She could see and appreciate, in the forests of several hundred dark and bright lines, the patterns that had led Miss Maury to develop her complex but coherent system. And yet, Mrs. Fleming's hydrogen-line alphabet also evinced logic, insight, and internal consistency. One approach focused on the overall pattern of spectral lines; the other emphasized the thickness or thinness of individual lines. Each arranged the stars in a different order. Miss Cannon stirred the two approaches in her mind as she parsed the starlight of the nether hemisphere.

WHEN CATHERINE BRUCE SAW the evidence from Peru of her telescope's prowess, she thanked Pickering "a thousand times— no, many thousand times, for each star on those truly extraordinary plates." She had not seen the actual glass plates, but rather photographic prints that Pickering made from them, as gifts to her. They portrayed Solon Bailey's bountiful stellar clusters, captured by the giant telescope's all-seeing eye. Miss Bruce pronounced them "most wonderful productions." It tickled her, she said, to be for once in the position of recipient, with Professor Pickering as the donor. Meanwhile her own donor activity continued unabated. Requests reached her from astronomers everywhere, and she responded as Pickering advised. To Max Wolf of Heidelberg, the man who had made the first discovery of an asteroid by photography—and named that body Brucia in her honor—she gave $10,000 for a new telescope. As a charter subscriber to George Ellery Hale's "magazine," the *Astrophysical Journal,* Miss Bruce provided the $1,000 required to set the struggling publishing venture on firm financial ground.

In 1897 Pickering approached Miss Bruce on behalf of the young

Astronomical Society of the Pacific, entreating her to endow a gold medal as special recognition for lifelong achievement by an individual researcher. In agreeing to establish such an award fund, Miss Bruce stipulated that her medal, like her grants, should go only to the truly deserving, irrespective of nationality. She also thought it would be a grand thing to see the medal bestowed someday on a woman, and added that possibility to the eligibility criteria: "citizens of any country, persons of either sex." As for the rest, she wished Pickering would "arrange the whole affair." She was past the age of eighty now, tired, and often ill. She relied more and more on her sister to keep up her astronomy correspondence.

Pickering, along with two other directors of U.S. observatories and three from Europe, submitted nominations for the first Bruce medalist. The astronomical society's board of directors settled easily on the foremost American astronomer, the already much-decorated dean of celestial mechanics, Simon Newcomb of the U.S. Nautical Almanac Office, whom they considered a philosopher as well as an astronomer and a mathematician. Newcomb had supervised recalculation and tabulation of orbital elements for all the planets, and also improved the efficacy of several astronomical formulas by arriving at new values for fundamental constants, which were adopted by institutions the world over. Miss Bruce, who traced her active support of astronomical research to Newcomb's 1888 article in the *Sidereal Messenger,* approved the choice. Newcomb had changed his view of future prospects in astronomy. Now that her grants supported two of his computing projects, he called on her whenever his travels took him to New York. "I like Newcomb decidedly," Miss Bruce confided to Pickering, "but I believe I like all the astronomers whom I know." She had become a magnet for them. Pickering remained her special favorite, however, and his Arequipa telescope her most generous gift.

Within the Bruce telescope's first few months of operation in Peru, it produced photographic charts of the entire southern sky. These images at once supplemented and augmented the existing southern star catalogues. The Uranometria Argentina of 1879, for example, listed the position and brightness of 7,756 stars to the limit of seventh magnitude. A single Bruce exposure of three hours' duration collected the light of as many as four hundred thousand stars, some as faint as fifteenth magnitude. Pickering offered "glass copies of our negatives" to all interested astronomers, to be used as source material for any number of important investigations.

The Bruce telescope saw, as no predecessor ever had before, into the hearts of the southern star clusters and nebulae that Bailey loved to explore. In the spring of 1897, he asked Pickering's official permission to pursue "the variable stars in clusters (or elsewhere) which I have or may discover." Bailey predicted his proposed study of such stars' periods might occupy his spare time for many years, during which he would always welcome the director's assistance and advice.

Pickering approved the plan, without foreseeing its potential as a source of friction between his two camps. Since Bailey had specified variable stars "which I have or may discover," he became increasingly concerned with their discovery. He and his assistants, DeLisle Stewart and William Clymer, began checking each night's images to pick out the possible variables before shipping the plates to Cambridge. Soon Mrs. Fleming complained.

"She feels," Pickering explained to Bailey on September 29, 1897, "that in these cases the credit goes to the Peruvian observers while a large amount of work falls upon her. She is obliged to measure the positions, the variations in brightness, if any, and to identify the individual lines, classify the object and see if it is a catalogue star. She also has to reexamine the plates since the fainter objects, including about half of the peculiar objects, and as many more having slight peculiarities are omitted [from the existing catalogues]. All of this is part of her regular routine work and has been for the past ten years, and much of it could not be done in Peru.

"On the other hand," Pickering conceded, "Dr. Stewart would doubtless feel aggrieved if after all the labor especially in following the Bruce plates, he is not allowed to examine them. The delay might also prevent the early discovery of a new star or other object of special interest."

Bailey, for his part, sympathized with Mrs. Fleming's vexation, but thought it unfair to deny credit for producing "first class plates," while publicly recognizing "the mere picking up of new objects by certain well known characteristics." Here Pickering had to agree, and he promised to change observatory policy accordingly. From now on, individual assistants who demonstrated special skill or care in photography would indeed receive proper acknowledgment in Harvard announcements.

Miss Maury had often feared she might not receive credit for the years of effort she invested in her classification system. But in 1897, when her "Spectra of Bright Stars" was published in the *Annals* of the Harvard College Observatory, "Antonia C. Maury" stood out in bold black and white, right on the title page, *above* the name of Edward C. Pickering, Director. It marked the first time that a woman had authored any part of the *Annals*. In 1890, in contrast, the contributions of "Mrs. M. Fleming" to "The Draper Catalogue of Stellar Spectra" had been described and gratefully acknowledged only in the director's introductory remarks.

Pickering's preface to the new volume noted that Miss Maury had been assigned in 1888 to study the spectra of the bright northern stars as part of the Henry Draper Memorial, "and she is alone responsible for the classification." Given that her investigations had been made several years prior, he said, they predated "the recent discoveries respecting the spectrum of helium." Rather than rewrite the lengthy treatise in the light of helium, Miss Maury had appended a discussion and some new thoughts in six pages of "Supplementary Notes."

After her return from Europe in 1895, Miss Maury retreated to the old Draper homestead in Hastings-on-Hudson, where her mother had grown up. The domes of her uncle Henry's telescopes now stood empty at the top of the hill, but the several cottages on the property still belonged to her elderly great-aunt, Dorothy Catherine Draper. Miss Maury found work in nearby Tarrytown-on-Hudson, teaching chemistry and physics at Miss C. E. Mason's Suburban School for Girls.

Nostalgia carried Miss Maury back to Vassar, also nearby in Poughkeepsie, for the astronomy department's annual dome party. In her college days she had studied under Maria Mitchell, the first lady of American astronomy, who instituted the dome parties and the practice of asking all the student guests to write poems on scraps of paper. Miss Maury felt inspired to revive that tradition. Her 1896 "Verses to the Vassar Dome" began, "A low-built tower and olden, / Dingy but dear to the sight, / And they that dwell therein are wont / To watch the stars at night."

The late Professor Mitchell had been just under thirty when she attained world fame and a gold medal from the king of Denmark for her 1847 discovery of "Miss Mitchell's Comet." Miss Maury had herself only recently turned thirty, but her career seemed to be veering away from astronomy. Perhaps her publication in the *Annals* would steer it back in the direction she intended years earlier, when "with searching glass I scanned / Those far deep lanes of night, / Where stars well up through endlessness, / In springs of living light."

Pickering invited Miss Maury briefly back to Harvard in mid-August 1898, to speak about her early research at a collegial meeting of distinguished astronomers intent on forming a national professional society. Everyone came, from Simon Newcomb, the elder statesman of the science, to thirty-year-old George Ellery Hale. Hale, who so successfully organized the country's first astronomy meeting in Chicago in the summer of 1893, had hosted another one in 1897, at the dedication of the grand new Yerkes Observatory in Williams Bay, Wisconsin, where he was now director. His arrival in Cambridge in 1898 coincided with a severe heat wave that lasted throughout the three-day gathering. Pickering's welcome proved equally warm. As the attendees were too numerous to be accommodated en masse in the observatory, the director ushered all hundred-plus of them into the parlor of his home.

"The spacious mansion of Professor Pickering formed an ideal place for the meeting of a convention," writer Harriet Richardson Donaghe reported in *Popular Astronomy,* "while the gracious dignity of the director and the hospitality of his stately wife, who received their guests with a cordial greeting, gave the serious purpose of the assemblage a touch of festivity and saved even the non-scientist from feeling out of his element." Miss Donaghe herself was one of the few nonscientists present. The situation called to her mind Walt Whitman's poem about the "Learn'd Astronomer," and she quoted part of a line from it in her article: "'Charts and diagrams, to add, divide and measure,' gave evidence of the heavy work laid out for the *savants,* but behind them gleamed the snowy outlines of the bust of some honored ancestor, the rich coloring of a family portrait, or the sparkle of a jewelled miniature, in the artistic setting of a private drawing room."

Members of Harvard's own observatory staff peopled the roster of speakers, beginning with Professor Searle, who delivered a talk on the "personal equation," or the way an individual observer's visual acuity, eye-hand coordination, and speed of reaction affected his perceptions. Mrs. Fleming prepared an announcement of the numerous new variables with bright hydrogen lines found on Bruce and Bache telescope plates from Arequipa. The director read her paper aloud at the podium, adding a coda of his own. As Miss Donaghe reported, "In conclusion Professor Pickering said that Mrs. Fleming had omitted to mention that of these seventy-nine stars nearly all had been discovered by herself, whereupon Mrs. Fleming was compelled by a spontaneous burst of applause to come forward and supplement the paper by responding to the questions elicited by it." Later Solon Bailey, only recently returned from his five-year tour of duty in Peru, discussed his pet subject, "Variable Stars in Clusters," and at the end Miss Maury held forth "On the K Lines of Beta Aurigae."

A coterie of participants, including Newcomb and Hale, met privately with Pickering to define the national astronomical society and draft its constitution. They did all that in a single day, though they could not settle on a name.

By the time the members of the nascent organization dispersed to their home institutions, the entire astronomy community knew that an important new solar system body had come to light in Europe. Gustav Witt of the Urania Observatory in Berlin and Auguste Charlois at Nice were both out hunting asteroids with telescope and camera when they picked up the object's trail on the night of August 13, 1898. Max Wolf's detection of Brucia in 1891 had proven the superiority of photography for such pursuits: On a plate exposed for two hours or more, a fast-moving asteroid would stand out as a very short line against the backdrop of the distant point-like stars. Upon developing their plates, both Witt and Charlois detected the same small streak of evidence, but Witt filed the first claim, with the result that other astronomers joining the chase in the following days referred to the quarry as "Witt's planet." The newfound body quickly distinguished itself as the speediest of its kind, and therefore bound to pass nearer to Earth than any of the others.

Seth Carlo Chandler, a specialist in defining the orbits of comets and asteroids, rushed to determine the true path of Witt's planet. Once he had worked out a preliminary ephemeris, or table of its predicted positions, from the current spate of sightings, he estimated that the object, now barely discernible at eleventh magnitude, probably had whistled close by Earth in 1894. No one noticed it then, but Chandler hoped a trace of its passage might be preserved on one or more photographic plates in the unique astronomical archive of the Harvard College Observatory. He would need to smooth over his earlier dispute with the director over the Harvard Photometry to gain access to the glass universe.

"I deem it my duty, in the interest of science," Chandler wrote to Pickering on November 3, 1898, "to send you the enclosed ephemeris of the planet . . . which all astronomers will be interested in, in common, for the recovery of any previous observations of this most important body." Pickering of course assented, and directed Mrs. Fleming to search through the plate stacks. With Chandler's rough map in hand, she selected the likeliest photographs among the hundred thousand in storage, and spent months combing through them for signs of Witt's planet. In early January 1899, on plates dated 1893, she finally found the elongated blob that she took to be the asteroid, and measured its positions. Chandler then incorporated the additional data into a corrected orbit, which he sent back to her. Armed with the improved map, Mrs. Fleming once again located the object, which had been named Eros in the interim, on plates from 1894 and 1896.

"I thought the asteroids were always feminine," Miss Bruce exclaimed when she heard the news. And it was true that all 432 previous discoveries (beginning with the first one, Ceres, in 1801) had been given women's names. "Fortunately poor dear little Eros is so far from the rest," she added, "or his life might be made miserable among all those old maids. It is pleasant that you photographed him so long ago while he was yet happy—unknown to fame."

Mrs. Draper wrote to say she was glad the ever-growing trove of photographs had cornered Eros, noting "the small god himself could scarcely have been more troublesome." Chandler agreed. He thought Pluto a more suitable name than Eros for the asteroid, "for malignity alone."

As Chandler redefined Eros's orbit around the Sun, he predicted the asteroid would come to pass very near Earth in the autumn of 1900. At that close range, Eros might be pressed to answer the oldest riddle in astronomy: What is the distance between Earth and the Sun?

The remoteness of the heavenly bodies made their distances all but impossible to gauge. The most the ancients could say was that the planets must be closer than the stars, since the planets, or "wanderers," could be seen moving with respect to the stars, while the constellations ever maintained their same configurations. In the third century BC, Aristarchus of Samos judged the relative remove of the Sun and the Moon by geometry, concluding that the Sun was probably twenty times farther away than the Moon.

In the 1500s, when Copernicus proposed the planets circled the Sun and not Earth, he estimated the relative distances between these bodies. Jupiter, for example, must lie 5.2 times farther than Earth from the Sun, and Venus only a fraction (0.72) of Earth's solar distance. But Copernicus still had no idea how far away to place the stars. Nor could Kepler, who derived the laws of planetary motion in the early 1600s, offer more than relative proportions for the distances separating the members of the solar system. Determining the true width of a single interplanetary gap would define all the others at a stroke. And a firm figure for the Earth-Sun distance would constitute a crucial milestone en route to the stars.

An opportunity to define the much desired Earth-Sun distance, or astronomical unit, arose late in the eighteenth century, on the occasion of the 1761 transit of Venus. Twice in about a hundred years, the orbits of Earth and Venus allow the sister planet to be seen crossing the face of the Sun over a period of several hours. English astronomer royal Edmond Halley foresaw the phenomenon's potential for resolving the distance dilemma. He imagined observers venturing far to the north and south of the globe to watch the transit and record the exact times of its various stages. The wide geographical separation between the observing parties would cause each to see Venus transit the Sun at a slightly different solar latitude. Later, by comparing their notes and triangulating, they could deduce the distance to Venus and extrapolate the Earth-Sun distance. "I wish them luck," Halley said in laying out his scheme, "and pray above all that they are not robbed of the hoped-for spectacle by the untimely gloom of a cloudy sky."

Clouds indeed intervened in some places to foil the observations. Even where clear weather prevailed, the hundreds of astronomers who heeded Halley's call failed to achieve precise measurements, so that neither the 1761 transit nor the next one in 1769 yielded the desired result. The great effort and expense did succeed, however, in narrowing the Earth-Sun distance to a range of possibilities somewhere between ninety and one hundred million miles.

When the transits predicted for 1874 and 1882 again united scientists in the quest for a conclusive determination, Simon Newcomb took charge of American expeditionary preparations. In the run-up to the events, he commissioned the firm of Alvan Clark to build instruments and invited Dr. Henry Draper to Washington to teach the several teams how to photograph the Sun. In the aftermath, in the 1890s, Newcomb asked Miss Bruce to pay salaries for a staff of computers to reduce the accumulated observations. This process was still in progress when Eros entered the scene, promising to shave many thousands of miles of uncertainty off the long-sought figure.

Planning the Eros campaign of 1900–1901 mobilized the world's astronomers, though not in the mounting of expeditions. No one needed to go anywhere. Unlike an eclipse or a transit, which transpired in a matter of minutes or hours, the autumn visit of Eros would occupy several months' worth of nights. Observatories throughout Europe, Africa, and across

America were already ideally situated, and already outfitted with the large telescopes required for sighting a dim, tiny asteroid against a starry backdrop. An international consortium of cooperating astronomers would monitor Eros's changing position in relation to large numbers of reference stars. In the United States, only the Harvard College Observatory was equipped to track Eros by means of photography.

The escalating enthusiasm for Eros made Miss Bruce wish that her own asteroid, Brucia, "would show herself again." But the time was not right. Miss Bruce's namesake hid far out of sight. Miss Bruce also withdrew. She fell ill again and died at her home in New York on March 13, 1900.

"It is no easy thing to choose fitting words to refer to the close of any life on Earth," *Popular Astronomy* editor William W. Payne wrote in his obituary notice, "much more is it difficult to offer a right and worthy tribute to the memory of one like Miss Catherine Wolfe Bruce, who, for noble cause, the world of science has learned to love for what she was and for what she did." Payne, whose own Goodsell Observatory at Carleton College in Northfield, Minnesota, had once received aid from Miss Bruce, praised "her intelligent generosity," which "knew no limits of race or country, and so science the world over mourns a common loss. Her kind and thoughtful care lightened many a burden in her own land, awakened new zeal in needful research, and helped to finish many a task when patience and other resources were nearly gone." In closing his brief sketch of her life, Payne itemized the long list of her gifts to astronomy. They totaled more than $175,000—the equivalent of a royal ransom.

PART TWO

Oh, Be A Fine Girl, Kiss Me!

It was almost as if the distant stars had really acquired speech, and were able to tell of their constitution and physical condition.

—Annie Jump Cannon (1863–1941)
Curator of Astronomical Photographs,
Harvard College Observatory

The fact that I was a girl never damaged my ambitions to be a pope or an emperor.

—Willa Cather (1873–1947)
Winner of the American Academy of
Arts and Letters Gold Medal for Fiction

CHAPTER SIX

Mrs. Fleming's Title

MINA FLEMING'S STAR was on the ascendant. In 1899, at Pickering's urging, the Harvard Corporation formally appointed her to a newly created position as curator of astronomical photographs. She thus became, at age forty-two, the first woman ever to hold a title at the observatory, or the college, or the university at large.

At this same time, the turning of the century inspired the Harvard administration to assemble a time capsule of campus life, with photographs, publications, essays, and diaries solicited from students, faculty, and staff. Mrs. Fleming dutifully wrote out her contribution for the "Chest of 1900" over a period of six weeks.

"In the Astrophotographic building of the Observatory," she began March 1, 1900, on a lined yellow notepad, "12 women, including myself, are engaged in the care of the photographs; identification, examination and measurement of them; reduction of these measurements, and preparation of results for the printer." Every day they bent to their examination tasks in pairs, one with a microscope or magnifying glass poised over a glass plate in its frame, and the other holding a logbook propped open on a desktop or in her lap, recording the spoken observations of her partner. A hum of numbers and letters, like conversations in code, pervaded the computing room.

"The measurements made with the meridian photometer," Mrs.

Fleming continued, "are also reduced and prepared for publication in this department of the Observatory." Florence Cushman, who had previously worked for a business firm, received the sheaves of magnitude measurements made nightly with photometers in Cambridge and Peru. She and Amy Jackson McKay copied over the visual observers' judgments, calculated the corrections, and checked and rechecked the figures before consigning them to the printer. The rest of the female computing staff, consisting of the sisters Anna and Louisa Winlock (daughters of the previous director) and the ladies who helped them process the data regarding star positions, remained in the west wing of the original observatory, as the Brick Building's limited space could not accommodate everyone.

"From day to day my duties at the Observatory are so nearly alike that there will be but little to describe outside ordinary routine work of measurement, examination of photographs, and work involved in the reduction of these observations." If Mrs. Fleming's days blended in sameness, as she claimed, they bore no resemblance to those of any other invited contributor to the Harvard time capsule. "My home life is necessarily different from that of other officers of the University since all housekeeping cares rest on me, in addition to those of providing the means to meet their expenses." She had to plan and purchase all provisions, plus give instructions to Marie Hegarty, the Irish maid she retained to clean house and cook the evening meal six nights a week. Although Mrs. Fleming was contracted to work seven hours a day at the observatory, she rarely arrived past 9 a.m. or left before six in the evening. "My son Edward, now a junior in the Mass. Inst. of Technology, knows little or nothing of the value of money and, therefore, has the idea but that everything should be forthcoming on demand." The frugal Mrs. Fleming minimized her expenses by inviting Annie Cannon to board with her on Upland Road. Miss Cannon proved companionable and came from a good family. Her father, Wilson Lee Cannon, was a bank director and former state senator in Delaware.

"The first part of this morning at the Observatory," Mrs. Fleming reported on March 1, "was devoted to the revision of Miss Cannon's work on the classification of the bright southern stars, which is now in preparation for the printer." Miss Cannon had picked up the knack for classification much faster than Mrs. Fleming expected. Of course, Miss Cannon enjoyed the advantage of college-level instruction in spectroscopy, as well as several years' experience as an assistant physics teacher and observer—opportunities that had been denied Mrs. Fleming. Still, there was no begrudging Miss Cannon the credit due her for making rapid, accurate evaluations of stellar types. She mirrored Miss Maury's ability to characterize individual lines in the hundreds of bright stellar spectra assigned to her, but she did not insist, as Miss Maury had done, on some altogether new scheme of her own devising. Instead Miss Cannon abided by Mrs. Fleming's lettered categories. She had in fact built a bridge between the two Harvard sorting systems by simplifying Miss Maury's double-tiered division and skewing Mrs. Fleming's alphabetical order. Since both those approaches were arbitrary, founded solely on the appearance of the spectra, Miss Cannon was free to assert her own sense of order. After all, astronomers could not yet tie any given traits of stars, such as temperature or age, to the various groupings of spectral lines. What they needed was a consistent classification—a holding pattern for the stars—that would facilitate fruitful future research. Miss Cannon thought it best to move Mrs. Fleming's O stars from the tail end to the top of the list, giving the helium lines precedence over the hydrogen, in the fashion of Miss Maury. B stars likewise ranked ahead of the A in Miss Cannon's appraisal. Beyond those rearrangements, alphabetical order again held sway, except where Miss Cannon conflated certain categories. C, D, E, and a few other class distinctions had fallen away. The resulting order wound up as *O, B, A, F, G, K, M.* (A wag at Princeton later made the string of letters memorable by the phrase "Oh, Be A Fine Girl, Kiss Me!")

Mrs. Fleming's March 1 journal entry continued with "the classification of the spectra of the faint stars for the Southern Draper Catalogue." This was Mrs. Fleming's own province, though she shared the vast territory with Louisa Wells, Mabel Stevens, Edith Gill, and Evelyn Leland. Whereas, at the beginning of Mrs. Fleming's career, the faint stars of the northern sky had belonged to her alone, the southern sky could not be managed single-handedly. The observing conditions at Arequipa, for one thing, coaxed many more faint stars out of the dark. On plates made with the Bruce telescope, even ninth-magnitude spectra appeared legible enough for the positions of individual lines to be measured. Moreover, any new-found variable necessitated a search through as many as one hundred previous plates of the same sky area, taken through the decade in Peru, in order to confirm the star's variability. Every year this part of Mrs. Fleming's work grew more laborious, owing to the ever-richer trove of material for comparison. The numerous discoveries that had brought her such pleasure, such acclaim—so many cuttings in her scrapbook—weighed on her now. Even the director admitted that it had become difficult to amass all the required data for one variable star before another turned up.

"The work of measurement is already well advanced," Mrs. Fleming said of the lines in the southern spectra, still on day one in her journal, "and we expect to accomplish much during the coming summer. Professor Bailey's observations with the meridian photometer in South America then came up for examination."

Solon Bailey, now back in Cambridge, was writing out the results of his five-year sojourn at Arequipa. His southern magnitudes, or brightness assessments, focused on the multitude of variable stars in star clusters—the "cluster variables," as he called them. Glass plates he had taken through the Bache, Boyden, and Bruce telescopes revealed some five hundred variables in those stellar agglomerations, and their photographic brightness needed to be rectified with his visual observations. Often he spent the night at the observatory, aiding the director in new observations or supervising one or another of the assistants. The Baileys' fifteen-year-old son, Irving, whose entire childhood education concerned the natural history and archaeology of the high Andes, now attended the Cambridge Latin School, in preparation for entering Harvard College.

"Various other pieces of work" claimed Mrs. Fleming's attention during that first morning of record, and in the afternoon several business matters called her to Boston. Later on, she wrote, "I joined Mrs. S. I. Bailey, Miss Anderson, and my sister Mrs. Mackie at the Castle Square Theatre. The play was 'The Firm of Girdlestone'* and we all enjoyed it. Mrs. Bailey tried to persuade me to stop over and dine with her, and spend the night, but my little family needs me at home in the morning. They are apt to be late for breakfast, and consequently for daily duties, when the head of the house is not there to get them going."

The next day at the observatory, March 2, Mrs. Fleming devoted herself "to miscellaneous odds and ends, and a gathering together of loose strands." These included keeping up with the scientific correspondence and sending out copies of the observatory's latest pamphlet, "Standards of Faint Stellar Magnitudes, No. 2," to all those affiliates, both amateur and professional, who followed the fluctuating brightness of the variable stars.

"Next in order came Miss Cannon's remarks on the classification of spectra. This is very trying work as so many things have to be taken into consideration, especially where it is found necessary to change the form of a remark." Each such comment for publication offered a specific, often lengthy description of some aspect of a spectral peculiarity. It took time to make Miss Cannon see "why we had changed 'one thing' and questioned 'another.'" Miss Cannon's remarks struck Mrs. Fleming as voluminous, threatening to fill a couple dozen double-columned pages of the smallest print. Not even Miss Maury had found it necessary to *remark* at such length.

The end of the day afforded Mrs. Fleming a quiet period for reflection. "My small family has deserted me this evening. I am the loadstone left to prevent the house from blowing away. After dinner Miss Cannon found that the clouds had cleared away and the stars were coming out, so she went over to the Observatory to get her observations of the circumpolar variables with the 6 inch telescope. Edward has gone down to study with Mr. Garrett who is in his section (Mining Engineering Course) in Tech. Neyle Fish, Edward's young friend who has been with us since Christmas night, has gone to make some calls, and I am awaiting Miss Cannon's return. If she gets home early we may be able to dispose of some of the

* Based on Arthur Conan Doyle's 1890 novel of the same name, *The Firm of Girdlestone* portrayed the deceitful dealings of a failing family-owned business.

questions regarding the remarks in her classification. Meanwhile I must see the 'Herald' and find out from it, if I can, the condition of the Boers and the British in South Africa. Edward talks of going out there when he finishes his course at the Institute."

Miss Cannon stayed very late at the telescope that night, which bumped the ongoing discussion of her remarks to the next day, March 3, a typical working Saturday at the observatory. Before lunch, Mrs. Fleming found time to examine a few southern spectrum plates. She lamented that supervision of routine procedures left her less and less time for the "particular investigations" that most interested her, or even "to get well settled down for my general classification of faint spectra for the New Draper Catalogue."

Saturday night guests chez Fleming amused themselves playing "India" (a form of rummy), "jackstraws" (pick-up sticks), and the board games "crokinole" and "cue ring." Sometimes a few friends sang for the rest of the company, but if not, there was plenty of pleasant conversation to go around. Mrs. Fleming prepared fudge and dates stuffed with peanuts to serve to a few guests, or, for a large soiree, creamed oysters with hot cocoa, cakes, and sweets. Cleaning up and winding down afterward with Edward and Miss Cannon, she might not get to bed till well past midnight.

"This is my day of rest and retirement so far as Observatory work is concerned," Mrs. Fleming noted Sunday morning, March 4, "but it brings my only opportunities for investigating the condition of household affairs, and I find the day all too short for them." The linens had to be changed, the family wash gathered for the laundress. "Alas! how matter of fact and different from the Sunday morning duties of other officers of the University."

TURN-OF-THE-CENTURY FERVOR found William H. Pickering plotting a new scientific adventure. He had recently burnished his international reputation by making a major discovery: In March 1899 he detected a new moon of Saturn orbiting beyond the planet's vast rings. The revelation of a ninth Saturnian satellite put William in a league with the observatory's exalted father-and-son former directors, the Bonds. Half a century earlier, in September 1848, William Cranch Bond and George Phillips Bond together had uncovered the eighth known moon of Saturn, which they named Hyperion. They had spotted it through the lens of the Great Refractor. William's new moon, like so many other recent findings at the observatory, turned up on Bruce telescope photographs. Although the object was extremely dim, below fifteenth magnitude, William flushed it from hiding by superposing glass negatives of long exposures taken on successive nights. Only one among the many minuscule grayish dots changed position from image to image. In keeping with the established nomenclature theme of mythological Titans for Saturn's attendants, William suggested the name Phoebe, and it stuck.

The possibility of another, even more significant discovery fanned William's desire to view the upcoming total solar eclipse on May 28, 1900, which promised to be visible across the southeastern United States. With the Sun's light masked by the Moon, and the favorable geometry of this particular eclipse, William hoped to discern a planet lying inside the orbit of Mercury. Several astronomers suspected the Sun of harboring a large close companion, or inter-Mercurial planet, and William believed his photography skills sufficient to expose the object. His brother, Edward, who usually frowned on the expense and possible futility of eclipse expeditions, blessed the plan. With the funds allotted, William was building a large camera capable of capturing a dim phantom under twilight conditions.

Mrs. Fleming anticipated joining the eclipse party. Examining the eclipse photographs for signs of a planet between the Sun and Mercury, though challenging, would probably not differ much from her morning task on March 5, when she relocated the missing asteroid Fortuna on four recent chart plates. Then, after critiquing some remarks by Professor Wendell, the director's photometry partner, on variable star magnitudes, she grappled again with Miss Cannon's remarks. "This takes more time and concentration of thought than any manuscript I have worked on since we put Miss Maury's volume (XXVIII, part 1) through the printer's hands," she judged. "If one could only go on and on with original work, looking for new stars, variables, classifying spectra and studying their peculiarities and changes, life would be a most beautiful dream; but you come down to its realities when you have to put all that is most interesting to you aside, in order to use most of your available time preparing the work of others for publication. However, 'Whatsoever thou puttest thy hand to, do it well.' I am more than contented to have such excellent opportunities for work in so many directions, and proud to be considered of any assistance to such a thoroughly capable Scientific man as our Director."

Throughout her journal, Mrs. Fleming expressed only positive sentiments about Edward Pickering, except on the matter of her remuneration. When she engaged him on March 12 in "some conversation" about pay, she got no satisfaction. "He seems to think that no work is too much or too hard for me, no matter what the responsibility or how long the hours. But let me raise the question of salary and I am immediately told that I receive an excellent salary as women's salaries stand. If he would only take some step to find out how much he is mistaken in regard to this he would learn a few facts that would open his eyes and set him thinking. Sometimes I feel tempted to give up and let him try some one else, or some of the men to do my work, in order to have him find out what he is getting for $1500 a year from me, compared with $2500 from some of the other [male] assistants. Does he ever think that I have a home to keep and a family to take care of as well as the men? But I suppose a woman has no claim to such comforts. And this is considered an enlightened age!"

For a week following that expression of frustration, Mrs. Fleming felt too tired in the evening to recap her long days. At first she thought this failure "due to laziness, which with me would be something heretofore unknown." It turned out to be the onset of the "grippe," which soon confined her to bed, weak and feverish. When her son came down with the same contagious complaint, their doctor put them both on a regimen of beef tea. As Marie, the domestic, also fell sick—too sick to tend them or even repair to her own home—the doctor found a temporary caretaker for all three patients.

Miss Cannon remained well and continued her nightly observation of the circumpolar variables. She divided her days between the glass plates in the Brick Building and the holdings of the observatory's library, where she engaged in new paperwork the director entrusted to her: She now had charge of maintaining a card catalogue of vital statistics on variable stars. This resource, begun in 1897 by a former assistant, already consisted of fifteen thousand cards listing every published reference to the approximately five hundred known variables, culled from bulletins, journals, and reports of observers all over the world. Miss Cannon could read both French and German, the other two languages of science. She fattened the decks of cards in the existing bibliography and created new card files as new variables came to light.

In mid-April, when Mrs. Fleming fully recovered her strength, and no longer needed to take a carriage to the observatory, she reviewed her time-capsule diary with a pang of contrition. "I find that on March 12 I have written at considerable length regarding my salary. I do not intend this to reflect on the Director's judgment, but feel that it is due to his lack of knowledge regarding the salaries received by women in responsible positions elsewhere. I am told that my services are very valuable to the Observatory, but when I compare the compensation with that received by women elsewhere, I feel that my work cannot be of much account."

EDWARD PICKERING HIGHLY VALUED the accomplishments and industry of Mrs. Fleming. Indeed, he planned to nominate her for the 1900 Bruce Medal. Who better? In view of the important part taken by women in American astronomy, he reasoned, and given the fact that the Bruce Medal had been established by a woman, it seemed only natural that the honor be bestowed on the woman who had made the greatest number of important astronomical discoveries to date, namely Mrs. W. P. Fleming of Harvard. The recent and much regretted passing of Miss Bruce, who had explicitly opened the award to women, seemed to underscore Pickering's argument, and he hoped the other members of the prize committee would agree to see things his way. Some resistance was to be expected, of course, just as the Harvard Corporation had resisted, for a time, his idea of conferring the "curator" title on Mrs. Fleming. The men of the corporation had similarly balked at his suggestion that they appoint Mrs. Draper to the observatory's Visiting Committee, but eventually they relented and she became its first female member.

Mrs. Draper's frequent visits to the observatory, with or without the excuse of a committee meeting, always cheered her. She loved seeing the Draper Memorial work in progress, and took interest in the other projects as well. In the spring of 1900, she expressed the desire to accompany the upcoming Harvard expedition to observe the May 28 eclipse. She had experienced only one total solar eclipse—that of 1878—but without actually *seeing* it, owing to her voluntary seclusion in the tent where she counted out the seconds of totality. This time she would have nothing to do but view the phenomenon in the pleasant company of her personal guests, Edward and Lizzie Pickering, along with Mrs. Fleming and Miss Cannon.

The director had not meant to join the eclipse party, as he was not needed for the planned observations and therefore could not justify the added expense. However, Mrs. Draper's magnanimous invitation changed his mind. She made all the travel arrangements—railroad tickets, boat accommodations, the hotel rooms in Norfolk and Savannah, even a book for Pickering to read on the way down, called *Confessions of a Thug.** "You will find it full of horrors," she promised. "It might make good reading for a steamer, it is sufficiently thrilling to keep one interested and is a page of history."

At the chosen viewing site in Washington, Georgia, the Harvard groups converged with astronomers from MIT and from Percival Lowell's Flagstaff observatory. The weather did not disappoint. William Pickering set up his special camera, which looked like a huge box, eleven feet long and seven wide, with four lenses of three-inch aperture arrayed to capture the inter-Mercurial panorama.

When the partial phase of the eclipse began, around half past noon,

* Originally published in 1839, this novel by Philip Meadows Taylor purported to be the true account of an assassin belonging to the Thuggee cult in India.

Mrs. Draper and the others avoided looking directly at the Sun, in order to protect their eyes from injury, but at the cry of "Totality!" about an hour later, they all looked up to drink in the sight.

Where the midday Sun had blazed, an eerie reversal now deepened the color of the sky and cast a sudden chill over the observers. The dark face of the new Moon hung like a great black hole overhead, surrounded by the glimmering fringe of the Sun's corona. The corona, invisible under normal conditions, extended its platinum-colored streamers as though reaching out to the planets Mercury and Venus, which now came into view against the background of crepuscular blue. The strange, beautiful vision commandeered the senses for a full minute. Then, as the Moon continued moving in its orbit, a blinding shaft of sunlight shot through a gap in the mountains on the lunar limb, signaling the end of the event.

"I shall always be glad that I have seen a total solar eclipse," Mrs. Draper wrote Pickering on May 30, 1900. "Simply as a spectacle it is magnificent and as the Chief Justice remarked 'it gave one a distinct thrill.'"

William's camera arrangement allowed him to make thirty-six plates of the eclipse. Unfortunately, none of the pictures yielded satisfaction, because someone inadvertently disturbed the instrument during the brief totality.

The Eros campaign—the global effort to observe the newly discovered asteroid—proved more successful for Harvard. The privileged position of the Bruce telescope in the Southern Hemisphere enabled DeLisle Stewart at Arequipa to take some excellent photographs a month before the asteroid became visible anywhere else. Officially, Pickering cooperated with some fifty observatories worldwide to ascertain positions of Eros in the attempt to derive the Earth-Sun distance. Personally, however, he found the asteroid's changing light even more intriguing. Viennese astronomer Egon von Oppolzer had shown Eros to be as variable in brightness as many a variable star, and Pickering hoped to construct its definitive light curve. He recalled that when Mrs. Fleming first retrieved the Eros plates, she had pointed out slight brightness variations in the asteroid's trail. At the time, he attributed the irregularity to patches of haze in the intervening air; now he recognized other possibilities. Eros might prove to be a rotating body, with surface features of dramatic contrast, or perhaps a pair of two small bodies of differing complexion, tumbling around each other. Beginning in July 1900, Pickering directed the Cambridge chief of photography, Edward Skinner King, to make plates of Eros every clear evening through the 8-inch Draper telescope. Inside the dome of the Great Refractor, Pickering himself gauged Eros's magnitude visually by comparing its varying brightness to the stars along its path.

PICKERING'S BID ON BEHALF OF MRS. FLEMING failed to win her the 1900 Bruce Medal from the Astronomical Society of the Pacific. In January 1901, however, he learned that he himself was to receive a medal—his *second* gold medal from the Royal Astronomical Society. The first one, in 1886, had recognized his exhaustive work on the Harvard Photometry, or "the comparative lustre of the stars," as his English admirers described it. The 1901 medal extolled his studies of variable stars and also his advances in astronomical photography. The U.S. ambassador to the United Kingdom, Joseph Hodges Choate, agreed to accept the medal for Pickering at the February 8 ceremony in London.

"I do not know when I have heard so much approval expressed about the award of a medal," Mrs. Draper chortled. "Everyone I meet, who understands the subject at all, is delighted at the compliment paid to you, and what amuses me very much, I am brought in for a share in the congratulations, which I do not merit; but I shine a little from reflected glory." In fact the RAS president, Edward B. Knobel, cited Mrs. Draper by name in his prize-presentation address. He saw her as the prime enabler of Pickering's award-winning research, and commended "her beautiful idea" to hallow her husband's memory by embracing, enlarging, and enriching the science in which Dr. Draper had labored.

President Knobel also took the opportunity to praise Mrs. Fleming, "that most careful observer" among Pickering's "lady assistants," distinguished for her many discoveries regarding variable stars and stars with peculiar spectra. He mentioned her name not once but three times in the same speech.

Mrs. Draper happened to be visiting London when Miss Cannon's classification came off the press in late March 1901. Pickering immediately shipped a copy to her, along with a typewritten note expressing his satisfaction with it.

Miss Cannon's classification not only unified the earlier work of Mrs. Fleming and Miss Maury, but also clarified the interrelationships among the several stellar categories. The whole population of stars now seemed to be distributed along a continuum according to their spectra. While many stars belonged unmistakably to one class or another, there were just as many that blurred the boundaries by sharing some characteristics of two neighboring types. Miss Cannon represented these overlaps with new numerical subdivisions. For example, she introduced the designation B 2 A for spectra displaying B-type strong Orion lines *as well as* a few of the pronounced hydrogen lines typical of the A class. Stars labeled B 3 A tended a little more markedly that way, B 5 A still more so, and B 8 A much more. Her system allowed for as many as ten steps between letters.

Miss Cannon thought her arrangement of the classification categories represented the stages of stellar development. Any given star would evolve from type O to type M in the course of its lifetime. Or perhaps progress moved the other way, from M to O. It was difficult to say.

Miss Maury's appreciation of the widths and borders of individual Fraunhofer lines had led her to institute subdivisions, which she labeled *a, b, c,* and *ac,* and which cut across her twenty-two categories. Miss Cannon missed none of these distinctions, but relegated her descriptions of the waviness or haziness of particular spectral lines to her "Remarks."

The publication of Miss Cannon's lengthy treatise gave Mrs. Fleming no reprieve from the tedious supervision of manuscripts. The backlog of the observatory's unpublished material, according to the director's estimate, would likely fill twenty-eight volumes of *Annals.* He charged Mrs. Fleming to focus her attention on that wealth of data, and "put it in shape for publication, or at least in such a form that its final publication would not be a matter of great difficulty." In consequence, the tally of her discoveries plummeted. Pickering commented on the fact in his annual report for 1901: "The number of objects with peculiar spectra found by Mrs. Fleming from an examination of the photographs is unusually small this year, as a large part of her time has been devoted to the preparation of the *Annals*."

In October Pickering repeated his endorsement of Mrs. Fleming for the Bruce Medal. But, once again, his entreaty failed to gain her that accolade.

A particularly virulent form of "grippe" felled Mrs. Fleming in November, forcing her to miss several weeks of work. Other staff members were similarly stricken that winter season, including Pickering in December.

"As you see," the director wrote to Mrs. Draper on January 10, 1902, "I have at last got possession of my type-writer so that I am now in communication with the world this morning for the first time. I am gaining strength every day, so that, except for observing, I can now do a great portion of my daily work. I have still, however, a great respect for those persons who can climb stairs without difficulty."

After wheezing his way up the steps to his office on the second floor of the Brick Building, Pickering could send photographic plates or messages down again via the dumbwaiter near his desk. Everything in the office stood near some part of the great round revolving desk that all but filled the room. It had been custom-built, eight feet in diameter, to provide as much surface area as a table twenty-five feet long and two feet wide. From his seat at the desk's perimeter, Pickering could easily reach the twelve-section rotatable bookcase at its center, or open any of the twelve drawers evenly spaced along the outer rim. Pickering's paperwork, laid out around the desktop in piles, orbited the bookcase. With a spin he could bring the draft of a journal article to hand, or the sheaf of letters for his signature, or the latest reports from Arequipa.

On the morning of February 1, 1902, Pickering arrived to find a present from Mrs. Draper awaiting him. It was a novelty clock for his office wall, along with a note congratulating him on his twenty-fifth anniversary as director of the observatory. A fête soon followed, organized by Mrs. Fleming. Toward 11 a.m. she and Mrs. Pickering summoned the director to the photographic library, where all the assistants gathered to offer their good wishes and gifts. The staff of the Henry Draper Memorial had chipped in on a comfortable desk chair; the other assistants gave him a silver loving cup a foot high. Pickering made a short speech, and then everyone shared a celebratory luncheon.

"I enjoyed it very greatly," he wrote later that day to Mrs. Draper, "much to my surprise, for you know I do not like occasions. No one could help enjoying the kind feelings they all expressed, not the least acceptable being those in your letter. Altogether it seemed to me to be a great success, and I shall always carry with me most delightful memories of it. I want to make my 50th anniversary 25 years hence a more formal affair. Will you help us receive our guests then? Please do not say that you have a previous engagement!"

In March, when Pickering found the quarterly expenses for the Henry Draper Memorial exceeding Mrs. Draper's allowance, he informed her that he would balance the books by drawing on contingency funds. But Mrs. Draper refused his offer. Her proprietary feelings for the project prevented her from allowing any moneys other than her own to support it.

"I have quite a sentiment about paying for it myself," she told him on March 30, 1902, "but I do not feel that at present I can increase the amount I have set aside for the work." She would rather see some part of the effort curtailed than relinquish her financial control. Pickering hastened to reassure her, both by letter and in person, that he wanted to manage the memorial exactly as she wished.

Meanwhile the aging, decaying wood of the overcrowded original structures clashed with the observatory's earned stature as one of the largest and most productive institutions of its kind. Pickering compared the mismatch to "that of a man with plenty of food who is dying of thirst, or who has no shelter in winter." With a $20,000 gift earmarked by an anonymous donor for physical plant improvements, Pickering added a plain brick wing to the Brick Building, thirty feet square and three stories high—big enough to hold another ten or fifteen years' accumulation of glass photographic plates. He also installed a hydrant on the observatory grounds, to augment the fire protection provided by chemical extinguishers and electric alarms. Still justifiably fearful of fire, Pickering mandated an observatory-wide fire drill every two months, in which all assistants and officers participated.

Toward the end of September 1902, when another quarterly accounting revealed the ongoing discrepancy between Mrs. Draper's payments of $10,000 per year and the cost of conducting the project at Harvard, she reiterated her concerns. "You will doubtless think it absurd that I should have any objection to being assisted from the funds of the Observatory, but I must confess to having a very strong feeling in regard to paying for this work myself. I hope, therefore, you will excuse my referring to the subject again, and not feel annoyed at my doing so." After all, the endeavor constituted her own loving monument to Henry. Although she had long before conceded the impossibility of carrying out his mission by herself, she remained resolute in her determination to fund it with her inheritance. She wished her current circumstances allowed her more leeway, but one of her nephews was selling his interest in the Palmer family estate, and she felt duty bound to purchase it from him, rather than allow outsiders to buy in.

Mrs. Draper knew she had every right to insist that the Draper Memorial budget be managed as she dictated, and yet, at the same time, she did not wish to appear unreasonable. On reconsideration, she agreed to let Pickering rely on supplementary support for the time being. "Whatever the indebtedness may amount to, I can return later," she promised, as much to reassure herself as to inform him of her plans. She would need to "sail very close to the wind," she said, through the coming winter, but, as soon as she settled the family business predicament, she expected "to feel quite easy again."

CHAPTER SEVEN

Pickering's "Harem"

DEMAND FOR COMPUTER POSITIONS at the Harvard Observatory ran so high that some young ladies with college degrees offered to work there for free—for a time, at least, until they could prove themselves worthy of hiring. Mrs. Fleming usually deflected these overeager applicants. Even if tempted to avail herself of short-term volunteer assistance, she did not consider it good policy to place the observatory under obligation to anyone for services rendered gratis.

Bona fide new job opportunities seldom opened, given the observatory's tight economy and the loyal longevity of its employees. Anna Winlock, for one, had been at her post even longer than the director, and by 1902 her younger sister, Louisa, was nearing two decades in the computing room. No one had left to get married since Nettie Farrar's departure at the start of the Draper project. As Mrs. Fleming could attest, the women now on staff were wedded to their work. No newcomers need apply. Then, all at once at the start of 1903, the director instructed her to hire ten new computers.

Funding for the sudden expansion came from a $2,500 grant awarded by the new Carnegie Institution of Washington. Pickering had applied for the support through proper channels, but he conveyed his thanks directly to Andrew Carnegie. Aware of the millionaire's commitment to building public libraries, Pickering described the plate collection in bookish terms.

"We have this great library of glass photographs," he wrote on February 3, 1903, "each unique, easily destroyed, and containing a vast number of facts relating to the entire sky, for some portions of which there have hitherto been no readers. This grant furnishes readers, who will extract from this storehouse of the history of worlds, facts heretofore unknown, and which, except for this collection, could never have been learned, since it contains the only record of them upon the Earth."

"Mr. Carnegie asks me to say that he is rejoicing at the receipt of your note," the steel industrialist's private secretary replied, "and he hopes the Carnegie Institution is to aid a hundred such things, and is to find, now and then, at long intervals, a man such as yourself to cooperate with."

The new "readers" at Harvard perused the chart plates (also called patrol plates) profiling each section of the sky. They traced the history of known objects and also of new objects as soon as any were discovered. They skimmed the nebulous regions for faint lights previously overlooked. They pored through star fields to recover "lost" asteroids that had gone missing for years.

In March, Pickering wrote again to his new patron with two pieces of news. "A few days ago, the Potsdam Observatory announced a new variable star, having the shortest period known. They had observed it carefully during the last nine months—our plates extend the work back to 1887. Last night notice came from the Moscow Observatory, of another interesting variable. They have 13 photographs of it,— we have more than 200, which without your grant we could not examine. Our library, like the Sibylline books,* is the only storehouse of these facts open to human knowledge. You have given us the key by which new facts regarding unknown worlds are daily revealed to us."

Four months passed before a personal reply arrived from Skibo Castle, the Carnegies' summer residence in Scotland. "My dear Professor, Go ahead; you are on the right lines. I hope to see you upon my return. I thought probably that I beat the record by selling 3 lbs. of steel for 2 cents, but the whole

* The Sibylline books contained the collected, rhymed wisdom of an ancient Greek oracle, transmitted to the Roman king Tarquinius Superbus by a prophetess.

constellation of Orion for a cent knocks me out. Take the cake! Come and see us when you are abroad."

MRS. DRAPER'S FINANCIAL SITUATION did not improve from 1902 to 1903. In fact, it worsened. For nearly twenty years, she had been channeling the income from one particular piece of her New York real estate into the Henry Draper Memorial fund, but in 1902 the city took over that property. "I have been seriously crippled in my income, by the loss," she told Pickering. By juggling other holdings, she had "succeeded in making ends meet" through the spring of 1903, but still she felt pinched. "I have not forgotten that I owe to the general fund of the Observatory nearly a thousand dollars, and I hope another year to be able to repay it."

Her anxiety prompted her to question the way her money was spent. Did all of it go toward the photographic study of stellar spectra? Or were the lines blurring between the observatory's several projects, possibly to her disadvantage? For example, she wondered to what extent her contribution funded the operation of the Bruce telescope at Arequipa, and how many of the photographs made with that instrument belonged to the Henry Draper Memorial, intended to examine stellar spectra. Moreover, she asked, was it wise to continue taking pictures "of the entire sky, night after night" in both hemispheres? She knew she had approved this pursuit, and even provided one of the instruments for it, but where would it all end? Had not the accumulated abundance of plates already grown unmanageable?

"I should be obliged," she wrote on June 15, "if you will give me a statement as to what the Observatory now possesses which you consider as belonging to the Memorial, including instruments, plates, printed matter, manuscript ETC."

Her questions staggered Pickering. Although his annual reports to the Visiting Committee discussed the progress of the observatory's various projects under separate subheadings, the work formed a grand unified edifice. Parallel lines of research met and entwined. Photographs of spectra led inevitably to discoveries of variable stars, which necessitated tracing changes of brightness backward and forward in time on stored images, which process revealed other objects of interest and suggested other studies. In sum, the spectra that Henry Draper had been first to capture on glass plates were now yielding not only the composition of the stars, as the doctor had once dreamed, but many other insights as well. Spectral evidence for motion in the line of sight, for example, had disclosed the speeds of many stars toward or away from the Sun. Pickering and Miss Maury had found spectral-line clues to the presence of two stars where only one had been previously known. Stars' relative temperatures could also be read in the spectrum, by the intensity of their radiation at different wavelengths. (Contrary to ordinary associations with the colors red and blue, the reddish stars were cool compared with those emitting mostly blue-white light.) The smooth continuum of spectral types—the way the Draper classification categories evolved gradually from one to the next—suggested that the stars themselves evolved, perhaps changing from type to type over the course of a stellar lifetime.

Pickering's reply to Mrs. Draper assured her that all the photographs taken with Draper instruments belonged to the Draper Memorial. "Of course," he reminded her, "each photograph becomes, like a book, a storehouse of information. It may thus be constantly consulted in future years. This is being done every day with many of the Draper chart photographs, at the expense of other funds. In the same way, in the study of the Draper Memorial of the variables discovered from their spectra, constant use is made of the great number of photographs taken with the Boyden, Bruce, and other instruments."

Pickering also stressed the dedication of the Draper Memorial employees. "You will be interested to hear that Mrs. Fleming, not contented with working all day at the Observatory, has undertaken to continue the preparation of the Southern Draper Catalogue in the evenings at her home. A measuring apparatus has been made for her and a recorder provided."

Mollified, Mrs. Draper wrote back, "I regret to hear that Mrs. Fleming is taking up night work. I appreciate her zeal and interest but fear she may over tax her strength and break down from over work— I should prefer to hear she intended to take a long vacation." Mrs. Draper herself was preparing to sail for Europe in July. Before leaving, she thought she should revise her will, so as to guarantee the permanent endurance of Henry's memorial.

"My purpose in providing funds for this Memorial has been, as you know, to perpetuate the name of Dr. Draper (my husband) in connection with original work in Astrophysics and especially in the photographic study of stellar spectra, and to contribute to the increase of knowledge in this department of Astronomy." But now she worried whether, "in the course of years," this line of work might be exhausted, and another avenue opened.

"In providing for the continuance of the work in my will, I have to remember that in a relatively short time, you and I will no longer be living, and I have to guard against the possibility that your successor may not be interested in this branch of research, and may prefer to use the fund in some other direction, if he can. In this he may be wise, but I am not disposed to trust his judgment alone, or that of the Trustees of Harvard University, which would be practically the same thing." She thought a committee of competent astronomers might be appointed to make appropriate decisions when the time came.

"I appreciate highly the aid which you have given me during the past 17 years," she reminded Pickering, "and the great amount of time and thought which you have bestowed on the work of this memorial. I believe its results are even now of the greatest interest, and whatever value they have is due to you."

AT NIGHT IN THE WEST WING, at the eyepiece of the 6-inch telescope, Miss Cannon judged the brightness of the variable stars in her keeping. She used the time-honored technique, fathered by variable star pioneer Friedrich Wilhelm Argelander, of comparing each variable to nearby stars that were either a little bit brighter or just a tad dimmer. The smaller the difference between the target and its neighbor, the better her estimate. It was useless to attempt direct comparison by eye between the ultrabright and the palest light, but the human retina could reliably gauge differences of one-tenth to one-half of a magnitude. Some of Miss Cannon's stars varied within that narrow range, and could be contrasted with the same comparison star at every stage. For target variables that changed over time by wider margins than a fraction of a magnitude, Miss Cannon recruited two or more neighbors to use as benchmarks. She gave each star a numerical code name tied to its location, and recorded all nuances in the accepted shorthand.

Miss Cannon was not alone in her solitary pursuit. A few yards away, out on the iron balcony surrounding the dome of the Great Refractor, the junior man, Leon Campbell, tracked his assigned variables through the portable 5-inch telescope, or sometimes a field glass, and often by eye alone. All over New England—all over the country, really, and in foreign countries as well—other variable star observers were engaged in the same occupation. Many amateur astronomers had been enchanted by one of Pickering's pamphlets, and acted on his suggestions as to which variables they should follow. At least once a month, when their local weather and the phase of the Moon permitted, the members of this volunteer army assessed their stars' brightness, gauging magnitudes by comparison just as Miss Cannon did, and sent their observations to her at Harvard. She knew some of the more serious contributors by name, such as Frank Evans Seagrave, who owned a private observatory in Providence, Rhode Island, and Mary Watson Whitney, professor of astronomy and director of the student observatory at Vassar.

Under Harvard's huge central dome, at the controls of the latest photometer, Oliver Wendell tracked the variables' slighter fluctuations—as small as three-hundredths of a magnitude. The director stood right beside him. Pickering remained devoted to these observations, and kept a count of the number of stellar assessments he made with each photometer he built. On the night of May 25, 1903, he recorded a personal milestone, his millionth photometric "setting," in the logbook. Pickering, afflicted with tuberculosis as a youth, had been warned against nighttime exposure at the start of his astronomical career, but by now he could boast he had discovered the fresh-air cure for consumption.

The stars, Pickering knew, were telegraphing important behavioral clues by their variations in magnitude. Just as the patterns of spectral lines revealed stellar chemical components, so, too, the range of brightness changes over time concealed underlying truths, the nature of which had yet to be grasped. For now, one had merely to track and record the changes, trusting that someday the myriad readings would yield to interpretation. Pickering was never one to speculate when there were data to accumulate.

Miss Cannon collected all the magnitude determinations made by co-workers and correspondents, and merged them with those of enthusiasts at overseas observatories, from Potsdam to Cape Town, who published their results in professional journals such as the *Astronomische Nachrichten* and the *Monthly Notices of the Royal Astronomical Society.* Since 1900, when she took charge of Harvard's variable star card catalogue, she had added twenty thousand new index cards. In 1903 she turned the whole unwieldy database into a series of tables that could be read by any interested party. Miss Cannon's opus, "A Provisional Catalogue of Variable Stars," appeared in the *Annals* and enjoyed immediate wide distribution.

Numerous other variable star catalogues, including three by Seth Carlo Chandler, had preceded Miss Cannon's, and yet she called hers "provisional." The term bowed to the quickening pace of ongoing discovery made possible by photography. A pre-photography catalogue, published in Vienna in 1865, had listed the 113 variables then known. Miss Cannon's volume contained 1,227. More than half of these (694) had been discovered on the Harvard glass plates: 509 in Solon Bailey's globular clusters of the Southern Hemisphere, and 166 by Mrs. Fleming, who spotted their signal bright hydrogen lines as she analyzed stellar spectra for the Henry Draper Memorial.

Miss Cannon's tables digested a vast quantity of information, from the position of each variable and its name or other designation to its maximum and minimum brightness, its period, and its spectral category in the Draper classification. One column specified the nature of each star's variability—whether it was a onetime wonder of the nova type, for example, or a regular repeater of short or long period. Here Miss Cannon relied on the system framed by Pickering in 1880, which divided variables into five varieties.

Like playing cards in a cosmic game of patience, the stars could be shuffled and dealt in various ways. One might sort them by "suit," so to speak, according to their spectra, or by the numbered "face value" of their brightness as expressed in magnitudes. The five types of variable stars could be represented by the picture cards—jacks, queens, kings, aces, and wild.

More than half of the thousand-plus stars in Miss Cannon's catalogue fell into Pickering's Type II, the long-period variables. These took up to a year, or even longer, to cycle through their variations. She could not fathom the causes of their slow fluctuations in magnitude—or of the more rapid rises and dips typical of Types I, III, and IV. Only the relatively rare Type V variables changed brightness for a known reason. These were "eclipsing binaries," meaning closely orbiting stars that took turns blocking each other's light. The Type V paradigm, called Algol, or "the ghoul" in the constellation Perseus, dropped from magnitude 2.1 to 3.5 every three days, when the dimmer member of the pair passed in front of the brighter one. The resultant partial eclipse lasted ten hours, after which Algol brightened again, right on schedule. Its regular and pronounced brightness changes, plain to the attentive naked eye, had attracted observers since the 1600s, and earned Algol the common nicknames of "winking star" and "demon star."

Mrs. Fleming, who liked to design and sew doll outfits for hospitals and fairs, portrayed Algol's dual personality as part of a series of astronomical dolls she crafted for the Arequipa families one Christmas, by pairing a large, male Algol figure with a miniature black Dinah doll. In May 1902 she discovered an Algol-type variable herself, while chasing a comet's path through the glass plates. Her find was the most recent entry in Miss Cannon's provisional catalogue. Unfortunately, Mrs. Fleming had no suitable spectrum with which to ascertain the new variable's Draper classification.

Although she had graded Algol itself B 8 A, and most of the other twenty-two Algol-type stars A, it was too soon to say where hers fit in the Harvard/Draper scheme. Miss Cannon was content to leave a blank space in the spectrum column alongside the latest Algol discovery. A good number of other blank spaces in her tables pointed up other lacunae, such as missing minimum values, uncertain periods, absent spectra, or questionable variable type. But that was the point of a provisional catalogue, was it not? To expose the gaps in knowledge?

HENRIETTA LEAVITT, THE RADCLIFFE ALUMNA and onetime assistant, came back to Cambridge in the fall of 1903. She had traveled twice to Europe and worked a few years as an art assistant at Beloit College in Wisconsin, near her family's current home, before realizing how much she missed the observatory. When she wrote to Pickering to say so, he offered her thirty cents an hour to return—a compromise between his regard for her abilities and the standard computer hourly rate of twenty-five cents. On these terms, she joined the new cadre of Carnegie-funded "readers."

Despite the earlier bonhomie of Mr. Carnegie, his eponymous institution abruptly ended its support of the Harvard Observatory in December 1903. With scant prospect for the grant's renewal, Mrs. Fleming had to dismiss the corps of newly trained assistants—all except for Miss Leavitt. Pickering freed other money to pay her salary as a full-time interpreter of texts in the plate library. For her first solo reading assignment, he gave her the Great Nebula in Orion.

The Orion Nebula, the central gem in the Hunter's sword, had been meticulously drawn and mapped by George Phillips Bond and famously photographed by Henry Draper. It remained a mysterious thicket of stars embedded in dark lanes of what appeared to be dust and gas. Recently Max Wolf of Heidelberg had studied the nebula and found it studded with variables. Someone needed to follow up Wolf's observations to confirm the variability of all those stars. Pickering believed he had the right someone in Miss Leavitt, as well as an incomparable collection of long exposures, some lasting several hours, to facilitate her search. Since Orion could be seen from both the Northern and Southern hemispheres, all the Harvard telescopes had photographed it, over a time span of more than ten years.

Miss Leavitt waded into the nebula, armed with an ingenious aid to brightness estimation—a little glass rectangle containing pictures of model stars at various magnitudes. This diminutive reference guide, about one inch by three, framed in metal and attached to a long handle, resembled nothing so much as a miniature fly swatter. Miss Leavitt liked to call it a fly spanker, since it was "too small to do a fly much damage." Within six months, she had confirmed sixteen of Wolf's variables and found more than fifty new ones, which were confirmed in turn by Mrs. Fleming.

Miss Leavitt achieved her next spate of discoveries by a different method. Edward King, the observatory's master photographer, made her a single positive plate from one of the many glass negatives of the Orion Nebula. On the glass positive, the stars shone white against a grainy gray background. Miss Leavitt superposed each negative on this positive, then examined the combination through a magnifying loupe. Unchanging stars tended to neutralize each other, but eight new variables popped out at her. Two months of this effort added another seventy-seven variables to Miss Leavitt's life list. She moved on to other nebulae in or near other constellations. She amassed another two hundred variable stars in the two nebulae seen by Ferdinand Magellan when he circumnavigated the globe in the 1520s. To Magellan's eye, they had resembled a pair of luminous clouds afloat in the southern night sky. Astronomers who later resolved the clouds into star clusters still called them by Magellan's name. In early 1905, in the Small Magellanic Cloud alone, Miss Leavitt uncovered nine hundred new variables.

"What a variable-star 'fiend' Miss Leavitt is," Charles Young of Princeton wrote in awe to Pickering on March 1, 1905. "One can't keep up with the roll of the new discoveries." Mrs. Draper expressed similar sentiments about "Miss Leavitt's remarkable discovery of variables," on March 11. As the tally continued to rise, Mrs. Draper wrote again in May to applaud "the large number of variables in the Small Magellan Cloud. It is certainly strange that so many of them should be found apparently close together. Will you please congratulate Miss Leavitt for me." She offered congratulations also to Pickering's brother, William, "upon the discovery of the tenth satellite of Saturn, [as] he is now proprietor of two of the planet's attendants."

William had waited four years for other astronomers to second his discovery of Phoebe, the ninth satellite of Saturn. By 1904 the little moon had been seen through several large telescopes other than the Bruce, and also proven to ply the most unusual orbit in the solar system. Phoebe circled Saturn in a backward, or retrograde, direction, against the traffic flow of the other moons and rings. This discovery led William to the inescapable conclusion that Phoebe had started life as an errant asteroid. When she wandered too close to Saturn, the giant planet had captured her and constrained her in a retrograde orbit.

William's success with Phoebe spurred him to examine more Bruce plates of Saturn's surroundings for signs of additional satellites. When he found what he believed to be the tenth moon, on April 28, 1905, he named it Themis, another Titaness of Greek mythology. He struggled with the calculation of its orbit, but none of the computers could be spared to assist him. As usual at the Harvard Observatory, the stars took precedence over the planets.

Edward Pickering kept fellow astronomers abreast of Miss Leavitt's advances by issuing a rapid-fire series of circulars. A few of these publications included small prints showing parts of the images she inspected. The prints, magnified for clarity and crowded with tens to hundreds of thousands of stars, conveyed the flavor of her Herculean task, which included approximating the percentage of variables per plate. "It is very difficult to count the faint stars which cloud the background," Pickering said with obvious understatement, "on account of their closeness to one another, and the number is certainly underestimated."

Occasionally Pickering described one of Miss Leavitt's finds as a nova or an Algol variable, but the great majority of her variables demonstrated only slight changes of about half a magnitude over very short time scales.

Perpetually aflutter, they cycled from maximum to minimum at least once every day. Their rapid fluctuations engendered a new photographic approach, involving several successive short exposures on the same plate, so that each star appeared as a series of dots. When Mrs. Draper visited for a couple of days, she professed herself more impressed than ever with the "vast work" accomplished. "You live so constantly in the midst of it," she wrote to Pickering on May 29, 1905, "that you can not so well appreciate it."

PICKERING REPRISED HIS LIBRARY METAPHOR in the fall of 1905, to deplore the underutilization of the plate stacks. The nearly two hundred thousand glass "volumes" drew only twenty readers to their contents, and the director was desperate for more. Harvard president Charles Eliot agreed to help. "I expect to be in Mr. Carnegie's house in New York on the 15th of November," he wrote Pickering from his office in the Yard, "and will then promote a visit by Mr. and Mrs. Carnegie to our Observatory, if I have the opportunity."

Mrs. Fleming also tried to revive the philanthropist's interest in the plate library. She sent a long letter to his wife, Louise Whitfield Carnegie, along with a small gift.

Mrs. Carnegie responded warmly from the family's vacation cottage in Fernandina, Florida, on January 11, 1906: "We left New York three days after Christmas and have been so occupied getting ourselves established here for the winter that I am only just able to turn my thoughts to the many kindnesses showered upon us at the Christmas season by our friends. Chief among these kindnesses is your own great kindness to us—The Wonderful 'Story of the Stars' in its dainty Christmas dress, with the remarkable lantern slides, form one of the most unique lovely greetings that can be imagined, & when we think of your kindness in taking all this thought for us, words fail me to express how deeply we are touched & how truly we appreciate this precious gift. Then too, may I add, when this gift is the original work of the great discoverer herself, are we not indeed a favored people and are we not blessed in knowing her? We are so proud of you as a Scotchwoman! & proud of your womanhood! For I believe the feminine mind is more capable of 'fathoming the Eternal Thought' just as the feminine heart nestles closer to Nature & to Nature's God."

After explaining the purpose of the Florida stay as a way to "woo back our little daughter to robust health," Mrs. Carnegie closed with the hope "that we may some day have the great pleasure of welcoming you, not only to our New York home, but better than that give you a 'Hieland' welcome to our home in bonnie Scotland." Mrs. Carnegie had been born in the Gramercy Park neighborhood of New York City, but regarded herself Scottish by association with Andrew, who had come from Dunfermline.

Margaret Carnegie, the couple's only child, was a precocious nine-year-old with a sprained ankle. Mrs. Fleming felt a bond with "Miss Margaret" that inspired her to send frequent tokens she thought the child might enjoy, including pictures of stars, a book that described the laying of transoceanic cables, and even a specimen of cable. "It gave me great pleasure to learn from your letter of February 16," she told Mrs. Carnegie, "that Miss Margaret was gaining in health and was enjoying herself in the sunshine and flowers of beautiful Florida. Also, that the valentines afforded her some amusement. On the evening of February 13 my brother and I spent the whole evening doing up and addressing Valentines for schoolmates and friends of his little fellows. Outside my scientific work, my next joy in life lies in giving pleasure to others."

Over the ensuing months Mrs. Fleming continued to unbosom herself to Mrs. Carnegie: "Of my own two boys, only one lived to grow up. He is 26 years of age and is one of President Pritchett's boys, having graduated from the Massachusetts Institute of Technology in 1901. He is a mining engineer, especially interested in copper, and has been with the Phelps Dodge Co. at their Copper Queen property in Douglas, Arizona, during the last year and a half. . . . He is a good boy and makes true friends wherever he goes, but being a boy, and in such a profession, he is away from me almost all the time. I am not in danger of being lonely, however, as I have my mother to look after, and she keeps us stirring." Mrs. Fleming's youngest brother, a recent widower, and his two sons, ages eight and twelve, also lived with her.

A newspaper announcement of the Carnegies' travel plans alerted Mrs. Fleming to the fact that their spring stay in New York would be brief. The news seemed to rule out the possibility of their visiting Cambridge to see the observatory, in which case, she said, she would "wait patiently for the fall," in the hope of welcoming them then.

Mrs. Fleming received perhaps the most pleasant shock of her life on May 11, 1906, just four days before her forty-ninth birthday, when the Royal Astronomical Society elected her to honorary membership. The society, formed in 1820, had recognized the role of women astronomers early in its history, by awarding a gold medal to Caroline Herschel in 1828 for her several discoveries of comets. No woman could be admitted as a full fellow, of course, but over the years the society had made honorary members of a few female British subjects, including, most recently in 1903, Lady Margaret Huggins, the wife of Sir William Huggins, Henry Draper's old rival. Mrs. Fleming became the first American woman to earn the distinction, or rather the first woman doing astronomical work in America. She was still a Scotswoman through and through, with the brogue to prove it. But, after her long and productive residency in the United States, she thought it might be high time she applied for American citizenship.

PICKERING HAD LIKENED THE PLATE COLLECTION to a library lacking readers. In 1906 the observatory's Visiting Committee compared it to a gold mine minus a refinery: "Like a mining company which has put out of the ground a vast quantity of precious ore, but lacks the means for reducing the ore and preparing the metal for market, the Observatory possesses a great store of knowledge in the rough, but wants the means for working this knowledge into useful shape for the benefit of the world."

Solon Bailey, still in Cambridge and elated by the burgeoning population of variable stars, urged Pickering in 1906 to unite the astronomers of the world in a new cooperation. Miss Leavitt had worked alone in a few isolated regions of space. Surely her phenomenal findings justified canvassing photographs of "the whole sky, in order to determine the number and distribution of all variable stars, to the faintest magnitudes possible." Without such a concerted approach, Bailey feared, any search would be plagued by needless duplication and wasted effort.

Pickering concurred. He issued another circular, inviting observatories everywhere to equip themselves with the proper telescopes and cameras "to study the distribution of the variable stars, and learn what part they play in the construction of the stellar universe."

Pickering estimated the sky held as many as fifty million stars brighter than sixteenth magnitude. He wanted to test the constancy of every last one of them. "The comparison of such a vast number of stars on several plates is indeed so great an undertaking, that at first it seems impossible," he allowed in his announcement. But the outcome he foresaw "would be a worthy achievement for the present generation of observers."

Given Harvard's position at the variable star vanguard, Pickering did not wait for others to heed his call. He pushed forward with his own research assistants and limited resources. He asked Miss Cannon and Miss Evelyn Leland to master Miss Leavitt's method. Then he divided the heavens, like Gaul, into three parts, and apportioned one to each of them.

LIZZIE PICKERING HAD FALLEN and broken her ankle in her bedroom in February 1903, and from that point on her health had gradually deteriorated. The break incapacitated her for more than six months; when she could walk again, she did not feel up to resuming her former activities. She did not even attend the party at Anna Draper's house in New York on December 29, 1905, to celebrate the twentieth anniversary of the Henry Draper Memorial. Mrs. Draper laid her friend's absence to a merely temporary illness, but upon visiting the Pickerings in March 1906, she could see the seriousness of the situation well enough. Afterward she wrote Pickering to say she hoped his wife would soon be better. What else could one hope?

In May, when Mrs. Draper desired yet another tour of progress at the observatory, she stayed at the Brunswick in Copley Square, and made plain she did not expect to be entertained. "Do not take the trouble to have luncheon for me, for I can take an early lunch at the hotel." Mrs. Draper also wrote directly to Mrs. Pickering, continuing their cordial correspondence as though it might carry on indefinitely. So many times over the years their thoughts had turned toward each other at the same moment, with the curious result that their letters often crossed in the mail.

Mrs. Pickering's surgery in June 1906 succeeded in easing her pain, but neither she nor her husband believed it would extend her life. They shaped their plans accordingly.

All that summer the director kept up his work at the observatory, just a few steps away from Mrs. Pickering's sickroom in the residence. In mid-August, when he learned that the English astronomers John and Mary Orr Evershed were en route to Cambridge, he cabled their ship to welcome them as graciously as possible under the circumstances. "The very serious illness of Mrs. Pickering may prevent our entertaining you at our house as, otherwise, we should wish to do." Mrs. Pickering died on August 29, and was buried in Mount Auburn Cemetery near the graves of her parents, Jared and Mary Sparks.

"My future interests in life are likely to be somewhat restricted, and my years of usefulness cannot be very numerous," Pickering predicted to Edwin P. Seaver of the Visiting Committee two weeks later. "The needs of the Observatory are so pressing, and so much could be accomplished by an immediate expenditure, that I am inclined to give it a large part of my savings." In September he paid the first installment of a promised $25,000 to be disbursed over the next three months, challenging other donors to contribute its equivalent.

Along with the many expressions of sympathy, Pickering received an appeal that autumn from Elizabeth Lidstone Bond, a granddaughter of the observatory's first director. Miss Bond apologized for intruding at such a time, but she and her sister required his advice on a personal matter. "You know, of course, of my Aunt Selina's poverty," she wrote on October 13. He did know of it, yes. Sudden poverty had driven Selina Cranch Bond to beg Pickering, almost the moment he took over the observatory in 1877, for employment. Although her father, William Cranch Bond, had provided for his heirs through the family's profitable watch, clock, and chronometer manufactory in Boston, a scurrilous trustee later cheated the descendants out of their inheritance. Pickering was still sending Miss Bond occasional computing work that she, now seventy-five, could do at home in Rockland, Maine.

Elizabeth and her sister, Catherine, the daughters of George Phillips Bond, had been impoverished by the same blow dealt their aunt Selina. They, too, had worked briefly for Pickering, as copyists and translators, before establishing themselves as schoolteachers. Elizabeth thought her aunt Selina, "the aged daughter of an eminent man like my grandfather, as poor as she is," might justly claim assistance from a particular Harvard pension fund. The sisters hoped Pickering could direct them to a member of the pension-fund committee, and counsel them about how to proceed, given "your position in the observatory, and the unfailing kindness and consideration you have always shown us."

Elizabeth conceded the many difficulties likely involved, not least her aunt's unwillingness to accept money from anyone. "She has made her independence her religion."

Pickering assured the anxious nieces of his aid, and wrote the same day to President Eliot to inquire about the pension fund. On learning that it could support former faculty members only, Pickering devised an annuity tailored to the situation. He would put $1,000 of his own, to be matched by sums from the Bond sisters and their cousins, into the Harvard Corporation for investment. Starting immediately, Selina Bond would receive $500 a year (nearly double her present salary as a part-time, at-home computer) for as long as she lived. In addition, she would be relieved of all responsibilities and granted the title of assistant emerita in the observatory, "in consideration of the distinguished and long continued service to astronomy of her father, her brother, and herself."

"Perhaps you can suggest some better arrangement," Pickering wrote to the Bond sisters. "In any case, she had better not know where the money comes from, but receive her first notification from the College Treasurer."

Elizabeth and Catherine Bond embraced the plan, except that they refused to let the director pay even so much as one dollar toward its execution. "May I add this moral?" Catherine asked Pickering in mid-November, when all the arrangements had been settled and her aunt surprised but gratified by the recognition conferred. "In your loneliness and sorrow it must be some consolation to you that in these past weeks you have been lifting such a weight of anxiety from the sister and daughters of your predecessor. So often my sister and I have been glad that you were our father's successor, but never more so than now!"

CHAPTER EIGHT

Lingua Franca

THROUGHOUT 1906, the triumvirate of variable star hunters—the Misses Cannon, Leavitt, and Leland—combed piecemeal through Harvard's photographic maps of the sky. Each woman's third of the heavens encompassed a score of subdivisions to be searched individually and repeatedly. Henrietta Leavitt, whether because of her greater previous experience or the luck of the draw in the sectors assigned to her, took an early lead. She found 93 new variables within months of the hunt's start in 1906, trailed by Annie Cannon with 31, and Evelyn Leland, 8. Competition for the most discoveries played no part in their pursuit, but Pickering tallied and reported the women's sums. Given the overall quest to learn the distribution of all types of stars in the universe, the absence of variables from a particular region provoked almost as much interest as a strong presence.

Pickering might have trisected the sky by giving Miss Cannon the north pole, Miss Leavitt the tropics, and Miss Leland the far south, where she was already well acclimated thanks to years of helping Professor Bailey sift the content of his star clusters. Instead of a gross division by celestial latitude, however, the director parceled out the fifty-five parts of Harvard's "map of the sky" as though dealing a hand of rummy. Thus Miss Cannon received section numbers 1, 4, 7, 10, and so on; Miss Leland numbers 2, 5, 8, 11, etc.; and Miss Leavitt the rest.

For each stellar region, each lady gathered plates in quintuplicate—four negatives, forming a time-lapse series, plus a positive print (white stars on dark background) from a fifth date, to serve as a basis for comparison. One by one, as Miss Leavitt had done in her exploration of the Orion Nebula, each negative was laid atop its corresponding print. Stars of constant brightness neutralized the differences between their positive and negative images. Variables showed their colors (in black and white) to the practiced eye.

The women marked all suspects for further investigation. Some of these panned out as truly new finds, others as familiar faces from previous dragnets. With more time or more womanpower, Pickering might have allotted more than five plates per area, but the efficient protocol made the most sense under current constraints. It even allowed him to approximate how many variables escaped detection. If, say, Miss Leavitt located ten in one section, nine of which proved to be new—never claimed by another observatory or captured in an earlier Harvard search—then many more undiscovered variables likely resided nearby. If, however, among ten identified, nine turned out to be already known, then very few others still lurked in that zone.

"Found two new variables," Miss Cannon noted in her diary on Saturday, February 23, 1907. "Went to the club. Very cold." Her club, the women's College Club of Boston, often drew her out for dinner and entertainment. The extreme cold persisted through Sunday. "Did not go to church."

After quitting the house on Upland Road to make room for Mrs. Fleming's mother, brother, and two nephews, Miss Cannon invited her widowed older half sister in Delaware, Ella Cannon Marshall, to come live with her in Cambridge. They spent almost every free evening and Sunday together, attending concerts and lectures, shopping, dining with friends, "pouring" at ladies' teas. Miss Cannon's Acousticon carbon hearing aid enabled her to enjoy all these things. Sometimes she brought Miss Leland or another coworker home from the observatory to lunch with "Sissie."

As Miss Cannon searched the deep space of the map plates for new variables, she continued her telescope observations and augmented her index card collection. Having twice updated her "Provisional Catalogue of Variable Stars" to append the new crops of 1903 and 1904, she hardly expected her "Second Catalogue of Variable Stars," published in 1907, to be her final word on the subject. Miss Cannon was a census taker in the midst of a population explosion. The Second Catalogue, though comprehensive, concentrated on the variable stars of long period. It did not include the multitude of short-period variables Miss Leavitt had uncovered in the Magellanic Clouds. Those required a separate treatment, currently nearing completion by Miss Leavitt herself.

"It may be asked," Solon Bailey wrote in an article for *Popular Science Monthly,* "why it is necessary, or even desirable, to go on indefinitely with the discovery of new variables." Aside from "the value of adding any new fact about the universe to the sum of human knowledge," he offered an astronomer's version of the mountaineer's "Because it's there." Only after a great number of variables had been discovered, he said, and their changeable natures closely observed, could the inquiry into the causes of variability commence.

As the search for new variables entered its productive second year, Pickering continued to press for physical plant improvements that would give the books and pamphlets of the library, still housed in the old wooden observatory building, the same fireproof protection now guaranteed to the glass plates. Recent efforts by the Visiting Committee had failed to raise sufficient funds. On March 4, 1907, as though to demonstrate the danger, fire broke out in Pickering's half-empty house. The flames threatened to engulf the director's residence and leap to the adjacent east wing of the observatory. Fortunately, the members of the observatory brigade, their skills sharpened through years of surprise practice drills, heeded the alarm signal and doused the outburst even before the municipal firefighters arrived.

MINA FLEMING SPOTTED nineteen new variables in 1907. She identified them the same way she always had, by the vagaries of their spectra, as opposed to hunting for them by superposing chart plates. Only later, in the wake of discovery, did she turn to the chart plates to verify her finds. Catching a star in different guises at different times—brighter here, fainter there—clinched its identity as a variable. Following the precise course of its changes over time, however, sometimes required ten or more nearby steady lights to bear witness. Ideally the brightest of these neighbors would shine brighter than the variable at maximum, the faintest one fainter than at minimum, and the differences between the intermediates would not exceed half a magnitude. In 1907 Mrs. Fleming published her modus operandi for choosing and evaluating such sequences of stellar standards. "A Photographic Study of Variable Stars" gave positions and magnitudes for the more than three thousand stars she had corralled to track the two-hundred-plus variables she discovered.

"Many astronomers are deservedly proud to have discovered one variable, and content to leave the arrangements for its observation to others," commented Herbert Hall Turner, Mrs. Fleming's transatlantic colleague in the Royal Astronomical Society, in reaction to this work; "the discovery of 222, and the care for their future on this scale, is an achievement bordering on the marvellous."

In each group of stars, Mrs. Fleming measured the variable's neighbors with a fly spanker, gave them alphabetical labels beginning with *a* for the brightest, and then calculated the brightness difference between *a* and *b*, *b* and *c*, and every other interval along the sequence. Then she judged the same set of stars, in the same way, on a second plate, a third, and a fourth. Although the pecking order of her sequences held fast, the size of the brightness intervals did not. Some of the plates had been taken with the 8-inch Draper telescope, others with the Bache. Differences between the telescopes, and also between the photographic emulsions on the plates, introduced inconsistencies. She got around them by averaging the four figures for each interval. These mean values gave her a set of stepping-stones from one end of each sequence to the other.

Assuming, for the moment, that every star labeled *a* equaled magnitude zero, she pegged the magnitudes of *b* onward by adding on the successive intervals. Next she nudged these interim values away from the arbitrary zero starting point by correlating them to visual magnitudes. The director and his assistants, both in Cambridge and in Arequipa, had repeatedly observed many of her comparison stars, and recorded their magnitudes. She pulled those numbers from published reports and paired them with hers. Subtracting the difference between the visual and the photographic for each star, she derived a mean difference for each sequence. In the final step, she added the mean difference to each star to arrive at its "adopted magnitude."

Mrs. Fleming identified herself on the title page of her "Photographic Study" as "Curator of Astronomical Photographs." Later, on her petition for U.S. citizenship, she shortened the title to "Astronomer," since the form provided only a small space for stating one's occupation. In another box, she crossed out "Wife," typed "Husband" in its place, and put "deceased" in parentheses alongside James Orr Fleming's name. As of September 9, 1907, she was officially an American.

Having established and disseminated her photographic standards, Mrs. Fleming began the slow work of applying them to her 222 variables. Many of these stars appeared on one hundred or more plates, and she intended to measure their magnitudes on every available image, in order to ascertain all 222 light curves. Along the way, or in the future, whenever the true magnitudes of her comparison stars became known, then the light curves of her variables could be adjusted accordingly.

Everything was relative in the realm of magnitude assessment. Mrs. Fleming's photographic standards hinged on Pickering's photometry, which hung in turn on decades of visual comparisons of one star to another. The great desideratum of "true" or "absolute" magnitude awaited the discovery of the distances to the stars and the dustiness of space: distance dimmed every sort of illumination; and stardust, if such a thing indeed littered the heavens, might obstruct the flow of starlight.

While Pickering praised Mrs. Fleming's "Photographic Study" as "the first large collection of sequences of comparison stars for studying variables photographically," he was in the midst of honing a lone stellar sequence to serve as a universal standard. Miss Leavitt contributed mightily to this effort. One day, Pickering anticipated, the great chain of forty or more stars constituting Harvard's "North Polar Sequence" would underpin all photographic magnitudes.

At sixty-one, Pickering could still rely on his own keen vision for visual photometry. He was about to mount a new round of visual assessments of faint stars, using his latest photometer and a 60-inch reflecting telescope acquired from the estate of the late English astronomer Andrew Ainslie Common. The director's "personal equation"—the way his eye connected to his brain and hands—naturally differed from the personal equations of his assistants, Wendell, Bailey, and Searle, and yet the decades of iteration ad nauseam had brought a gratifying consistency to their results. The Revised Harvard Photometry, published in 1908, made this manifest. It provided cumulative data on the magnitudes of nine thousand bright stars. Pickering hoped that astronomers everywhere would respect this compendium of his efforts since 1879 as the standard reference authority in the field.

In recognition of all Pickering had done to further photometry and spectroscopy, the Astronomical Society of the Pacific awarded him its 1908 Catherine Wolfe Bruce Gold Medal for lifetime achievement. Pickering might have felt even more pleased to see the honor bestowed on Mrs. Fleming, as he had often suggested, but her prospects of winning it seemed unlikely.

THE REVISED HARVARD PHOTOMETRY, perused round the world, not only gathered and averaged information that had been scattered through several earlier volumes of the *Annals*, but also included the spectral type of each of the nine thousand stars, adjudged by Miss Cannon according to the Draper classification system. This useful addition soon elicited a criticism from a young Danish astronomer, Ejnar Hertzsprung of Copenhagen.

Hertzsprung shared Pickering's ardor for hands-on photometry. For several years he had been trying to factor distance into the stellar magnitude equation, in order to determine stars' intrinsic brightness. A number of stellar distances had been established by trigonometry for stars within one hundred light-years of the Sun. Relative distances of the farther stars could be wrung from their incremental movements, over time, across the line of sight, with the very farthest ones exhibiting the least so-called proper motion. By this yardstick, Hertzsprung revealed that some of the brightest stars lay at the greatest distances from the Sun. He could only imagine what blazing giants they must be, to beam such inordinate luminosity from the very depths of space.

In the spectra of the brightest faraway lights, Hertzsprung found very narrow, very sharply defined hydrogen lines. He recognized these traits as distinctions originally defined by Antonia Maury in describing the c division of her complex, two-tiered classification system.

As one of the first to see the wisdom of Miss Maury's ways, Hertzsprung rued the use of Miss Cannon's modified classification in the Revised Harvard Photometry. On July 22, 1908, he wrote to Pickering, complaining that the system adopted in the new volume was too simplistic. He compared it to a botanical classification based on the size and color of flowers instead of the morphology of plants. For emphasis, he reiterated the point with an animal analogy: "To neglect the c-properties in classifying stellar spectra, I think, is nearly the same thing as if the zoologist, who has detected the deciding differences between a whale and a fish, would continue classifying them together."

Pickering, the original publisher of Miss Maury's classification, appreciated its merits even while questioning its complexity. But Miss Maury had built her system on only a few hundred stars, and it might not hold true across tens of thousands. Likewise the conclusions that Hertzsprung drew from her work seemed to Pickering premature.

Miss Maury, who had never severed relations with Pickering, also wrote to him in mid-1908, to request another letter of reference. She was considering applying somewhere for an adjunct professorship in physics and astronomy. Without hesitation, the director again praised her "painstaking" investigations and "important" contributions. Presently she told him she would rather resume her research than pursue teaching. He assured her the observatory door stood open, though he could not promise her a full-time wage.

Miss Maury had long supplemented her income by giving freelance lectures, which she called "Evenings with the Stars." Her promotional brochure boasted she had spoken at Cornell University, Wells College, the Brooklyn Institute of Arts and Sciences, the New York Academy of Science in the Museum of Natural History, and the New York City Department of Education, as well as to schools, lyceums, clubs, and parlor audiences. Her terms were ten dollars for a single talk, "Sun, Moon and Stars—A Brief Survey," and thirty dollars for a four-part course on either "The Visible Universe" or "Evolution in the Heavens." She illustrated the presentations with lantern slides she requested of Pickering and Mrs. Fleming, who also sent her the observatory's circulars and other publications to keep her informed of scientific news through the years while she taught literature at girls' schools in the towns near Hastings-on-Hudson.

In December 1908 Miss Maury returned to the observatory as an associate researcher. She reunited with the spectroscopic binaries that had made her reputation nearly twenty years before, and also with Beta Lyrae, the mystery variable that changed its light on an irregular, inscrutable timetable. Miss Leavitt, similarly intrigued by Beta Lyrae's strange behavior, more than once opined to Miss Maury, "We shall never understand it until we find a way to send up a net and *fetch the thing down!*"

Miss Leavitt discovered another fifty-six new variables on the Harvard maps of the sky in 1908, maintaining her lead over Miss Cannon and Miss Leland by a wide margin. She also published her findings about the Magellanic Clouds. Through careful comparisons of many plates, she had observed the range of maximum to minimum brightness for all 1,777 of her variables, and listed these data in twelve dense pages of tables. Thus far, however, she had followed only a small number through their complete cycles of change. When she tabulated the periods alongside the magnitude ranges for these sixteen stars, a pattern emerged. "It is worthy of notice," she wrote in her report, that "the brighter variables have the longer periods."

She wondered what that might mean, and whether the trend would persist. She was continuing to analyze more periods when illness interrupted her work about two weeks before Christmas. From her hospital room in Boston on December 20, she thanked Pickering for the pink roses and get-well wishes. Then she went home to Wisconsin to recuperate.

The ideal weather conditions that smiled on Arequipa from May to October sometimes lasted all year long. Astronomers arriving from the north would remark on the stillness of the air that produced such optimal seeing. No radical day-to-night temperature changes unsteadied the dry atmosphere, and no predawn dew accumulated on telescope lenses. Those who stayed for many months almost welcomed the interruption of the brief cloudy season, which afforded them the time to make instrument repairs or tend to other neglected business. Of late, however, the off-season had lengthened. Cloud cover now spread a pall over the telescopes for extended periods between November and April. Harvard's staff members at the Boyden Station had persisted in their observations through revolutionary gunfire and local epidemics of smallpox and yellow fever, but clouds could not be endured. Pickering got busy soliciting opinions about alternate sites in South Africa. As before in Colorado, California, and Peru, he needed someone to reconnoiter the potential locations. Once again he chose Solon Bailey.

At a farewell luncheon, President Eliot saluted Bailey as the observatory's "foreign ambassador without portfolio." Fortunately, the new venture would not require the fifty-four-year-old emissary to climb mountains or build roads. His entire African itinerary stretched across a plateau with an average elevation of 5,000 feet. Though only knee-high to the Andes, the Cape Colony's Great Karroo plateau lay much farther south, and might facilitate the assembly of a South Polar Sequence to complement Pickering's northern one.

Bailey left Cambridge for Africa by way of England on November 17, 1908, traveling solo, with two telescopes, a camera, and various meteorological equipment in tow. On the advice gained in London from Sir David Gill and Sir William Morris, both veterans of long service to the Royal Observatory at the Cape of Good Hope, Bailey planned to set up a principal station in Hanover. From that base, he could make excursions to test regions of the Orange River Colony, the Transvaal, and Rhodesia.

Bailey rode five hundred miles from Cape Town on a train that deposited him at Hanover Junction shortly after midnight, then covered the remaining nine miles wedged into the backseat of a horse-drawn, two-wheeled wagon called a Cape cart. He arrived at the only hotel in Hanover by 2 a.m. "The driver opened a door on the porch, lighted a candle and left me." Bailey chose one of the room's two beds. "On the following day the proprietor and his wife appeared and did all in their power to make my stay comfortable."

In a ruled notebook with a blue moiré cover, Bailey graded the transparency of the African sky across a wide area over the course of a year. "The amount of clear sky and especially its distribution throughout the year," he reported, "are much more favorable than at Arequipa." On the other hand, the seeing, or steadiness of the atmosphere, was no better. Moving currents of air often made the stars jiggle in the telescope view. Temperatures rose and fell over a greater range than in Peru. There was more dew to contend with, not to mention the common occurrence of dust storms and violent thunderstorms.

"The vast stretches of plain, generally known as *veldt,* are parched and apparently dead in the dry season, but are often green and beautiful in the rainy season," Bailey discovered. "Each farm must have its natural spring (*fontein*) of water for domestic and farm purposes." Of all the places he surveyed in Africa, Bailey favored Bloemfontein, the capital of the Orange River Colony, as the best site for a permanent observatory. Its sky scored high grades for clarity on his scale, and the area "had much to offer in the way of social and educational advantages."

While Bailey was abroad, the sky conditions at Arequipa deteriorated further. Smoke issued from the long-dormant El Misti volcano, and also the Ubinas volcano erupted, about forty miles east of the station. As a further aggravation, the condition of Mrs. Draper's finances precipitated a drastic cut in funding.

"I have recently had to consider, very carefully, my financial condition and prospects," she wrote to Pickering on January 24, 1909, "and find to my great regret, that I shall be unable to continue to furnish, to the Observatory, the amount I have been giving, for the past twenty-three years, for the carrying on of the work of the 'Henry Draper Memorial.'" She set August 1 as the date she planned to reduce her monthly support to $400, less than half the accustomed sum. "I am exceedingly sorry to have to take this step, which will cause you as much surprise as it does regret to me—Fortunately, I believe that the special work for which I commenced my contribution, namely: cataloguing the stars by their spectra, is now in a fairly satisfactory state of completeness."

Mrs. Draper's generosity toward stellar spectroscopy had truly honored her husband's name. But its "completeness" opened new avenues for further work. Just recently the 11-inch Draper telescope, the same one that furnished Miss Maury with detailed spectra of the bright northern stars, had been turned on fainter stars, opening them to more intense scrutiny and possible refinements in their classification.

"I have hesitated as long as is wise about making this change, but find now there is no alternative," Mrs. Draper concluded. "I am happy that the necessity for reducing the allowance did not arise sooner, and that so much of value has been accomplished." In this, her seventieth year, she was appreciating the retrospective view.

Pickering, now sixty-two, abandoned all hope of undertaking an expensive transatlantic relocation of the Boyden Station. At Mrs. Draper's request, he summarized the results obtained to date with the Henry Draper fund, and projected how it would be used henceforward. He delivered the report in person.

"Since you were here I have more carefully looked over the paper," she wrote on February 14, the anniversary of their alliance, "and feel, as I told you, that we had every reason to congratulate ourselves." She regretted that her reduced payments would slow progress, but showed no diminution of her interest in the work or her affection for Pickering. "I enjoyed your little visit, so very much—It is always a pleasure to hear you talk of what is being accomplished at the Observatory. I wish you might run away more frequently."

MISS LEAVITT'S CONVALESCENCE at her parents' home in Beloit lasted longer than a year. When she finally felt ready to return to work, in January 1910, she was still not strong enough to travel to Cambridge. Pickering agreed to let her work remotely on determining magnitudes for the stars of the North Polar Sequence. Under the special circumstances, he sent her a set of glass plates with all the accoutrements she needed—a wooden viewing frame, a magnifying eyepiece, a ledger. She worked only two to three hours a day at first, but increased her effort as her strength returned. In May she reappeared at the observatory in good health, and boarded again with the family of her uncle Erasmus Darwin Leavitt, the mechanical engineer and inventor, who lived in a large house on Garden Street near the observatory.

The summer of 1910 brought a cadre of some twenty foreign astronomers to Cambridge. Luminaries included Astronomer Royal Frank Watson Dyson, representing both Edinburgh and Greenwich; Oskar Backlund from the Pulkovo Observatory in Russia; and Karl Schwarzschild, director of the Astrophysical Observatory at Potsdam. All of them had been invited to the United States by astronomical impresario George Ellery Hale.

Hale, now founding director of the Mount Wilson Solar Observatory in California, had helped create the Astronomical and Astrophysical Society of America in 1898, and later conceived a global organization to unite investigators devoted to his own chosen specialty, the Sun. The International Union for Cooperation in Solar Research, or the "Solar Union," met at Hale's instigation in Oxford, England, in 1905, and in Paris in 1907. In preparation for a Pasadena meeting in 1910, Hale hoped to enlist Pickering as a member. The influential Pickering, Hale thought, could help widen the reach of the Solar Union to encompass stars beyond the Sun. Furthermore

Pickering, as president of the Astronomical and Astrophysical Society of America, was perfectly positioned to host an East Coast open meeting of that organization, so timed as to bolster foreign attendance at the West Coast assembly of the Solar Union. Pickering agreed to gather the society's members and overseas guests at Harvard in August, and then chaperone the visitors cross-country by rail for the Union get-together on Mount Wilson.

"My adventures began before the train left Boston," Pickering wrote in his travel diary on Saturday, August 20, 1910. "The porters could not tell me in which car was my drawing room, and in passing from one car to another I got locked in between them! As the [William] Pickerings and Professor Bailey waved me a cheerful farewell, they did not know that I was a prisoner, alone in a glass cell, from which I could not escape."

Everyone had benefited from the talks at the three-day Harvard meeting just completed. Six of the distinguished foreign visitors had been elected, at their express request, to membership in the society. All enjoyed the way Pickering interspersed the technical sessions with appropriate diversions, such as the group excursion to the Harvard-affiliated Blue Hill Meteorological Observatory in Milton on Wednesday afternoon, and also the trip to the Whitin Observatory at Wellesley College on Thursday. On Friday, mindful of everyone's fatigue, Pickering moved the body only as far as the Harvard Students' Astronomical Laboratory in the Yard. Throughout the week, at all hours, his staff had ushered interested visitors to any on-site area they wished to see, from the telescope domes to the astrophotographical library in the Brick Building. Pickering wrote in his diary that he thought he might sleep for three days on the train West, but in fact he had a full agenda of important committee work to do en route.

Pickering's expertise in photography and photometry allied him with two major European stellar mapping projects, one headquartered in Paris and the other in Groningen. The moment had come for each of these projects to select a standard reference for photographic magnitudes. Pickering wanted to see a single standard applied to both efforts, and he wanted that standard to be the Revised Harvard Photometry. As chairman of the

Committee on Photographic Magnitudes of the Astrographic Chart Conference (the Parisian effort, also known as the Carte du Ciel), Pickering held considerable power, but other photometric standards had been devised, and the issue was to be decided by vote. Harvard's chief competition came from Pickering's fellow committee member Karl Schwarzschild, who had authored his own Potsdam standards of photographic photometry. As it happened, Schwarzschild was on the train with Pickering. So were committee members Herbert Hall Turner of Oxford and Oskar Backlund of Pulkovo, making a quorum. Indeed, the entire contingent of traveling astronomers was conveniently confined to two specially hired railcars.

On Sunday, August 21, 1910, they reached "Niagara, in which no one is disappointed," Pickering wrote. "The roar of the Falls, interrupted only by astronomical talk. Informal committee meetings whenever I sit still. In the morning a carriage ride to Goat Island, and the wonderful electric railway from which you see the entire river. In the afternoon the Maid of the Mist (steamer, not young lady) and an impressive view of the American Falls, from its base (the best I have seen). My coat got so wet that I had to keep my back to the Sun, to dry it."

They got to Chicago on Monday, toured the parks and the university physics laboratories, and were joined by several more astronomers when they reboarded the train at nightfall. John Stanley Plaskett of the Dominion Observatory in Ottawa, who made his own record of the trip highlights, relished the way the group "traveled across the continent in two special cars and in the eight days occupied in the journey became almost like a family party."

On Tuesday, August 23, after a long morning talk with Turner, Pickering called a meeting of the photographic magnitudes committee in his drawing room. "Backlund, Schwarzschild, Turner, and I discussed the matter for two hours, so busily that we did not find out that it was extremely hot until the meeting was over. Temperature 102 in the shade. The thermometer went down when you put the bulb in your mouth! At an open window the breeze was hot like that from a furnace. We all suffered, and several ladies were ill." Many of the visiting astronomers were accompanied by their wives, and Harvard's own well-known Mrs. Fleming was also aboard the train.

The next day, Wednesday the twenty-fourth, Pickering worked all morning to complete his part of the committee report on photographic magnitudes before another rolling meeting convened at three. This one included an additional member, Edwin Brant Frost of the Yerkes Observatory. Pickering re-created the drama later in his journal, in the present tense: "They don't want to come, as the thermometer is nearly 100, and point to Turner, who is asleep. I wake him up, and make them all attend the meeting in my drawing room. It is so hot they cannot contribute their portions of the report. As a result of our labors (and heavy labors, too) we all agree on a system of photographic magnitudes which will probably be the system of the world. I am repaid for my journey of two thousand miles, had I done nothing else. Astronomers very kind and complimentary, and Schwarzschild gives up his (Potsdam) system and accepts that of Harvard. My part in this will be regarded as one of the most important things I have ever done." Thus the acceptance of the Harvard Photometry standard, one of Pickering's top goals for the trip, became a fait accompli before the train crossed the Great Divide.

In Flagstaff, Arizona, on Thursday, Percival and Constance Lowell guided the visiting astronomers through the Lowell Observatory, then waved them on to the scenic wonder of the American West: "Saturday, August 27. Walk in morning to another point on the rim of the Grand Canyon. Prepare six copies of a third draft of the report on photographic magnitudes with the help of the hotel typewriter. Leave in evening for Pasadena." Pickering, flush with his success at advancing the Harvard photographic standards for photometry, could only hope that the Draper system of stellar classification by spectra would fare as well in the imminent contest for international approval.

Over the five decades since Father Secchi grouped the stars visually by color and a few spectral lines, classification schemes had proliferated. Harvard alone had produced two—or three, depending on how one counted

Miss Cannon's modifications of Mrs. Fleming's original Draper catalogue. A veritable babel of terminology prevailed. To make himself perfectly clear when addressing other astronomers, Pickering often translated Harvard designations into the simpler Secchi names, describing one of Miss Cannon's F 5 G stars, for example, as belonging to Secchi's second class (a spectrum crowded with many lines). Secchi's system, familiar as a Latin grammar, lacked the vocabulary to describe all the spectral distinctions revealed through photography and modern analytic techniques. Astronomers knew they could improve their communications by choosing one classification system to abide by, or else by creating a new hybrid. The issue was due to arise at Mount Wilson, when the Solar Union debated broadening its mandate to include other stars.

Wilted from the heat of the Mojave Desert, the astronomers reached Pasadena late Sunday afternoon, August 28, 1910, and checked into the Hotel Maryland. The collegial body that had banded together in Boston, then expanded in Chicago, now melded with West Coast residents and Solar Union delegates newly arrived from as far away as Japan. Eighty-seven strong, the attendees represented thirteen countries and fifty observatories at the largest gathering of astronomers ever assembled.

"Monday, August 29. The fourth anniversary." No doubt Pickering would continue to mark the date of Lizzie's death until the day of his own. This year he passed the grim occasion in good company, visiting the offices, laboratory, and machine shop of the Mount Wilson Solar Observatory. The facilities occupied a one-story concrete building in town, where Hale joined the group to describe the fabrication of the unique instruments they would see over the next few days on his mountaintop. At the afternoon garden party given by Hale and his wife, Evelina Conklin Hale, the astronomers met some of Pasadena's most influential citizens.

It took all of Tuesday to reach the summit of Mount Wilson. A few of the astronomers, though dressed in suits, ties, and derby hats, mounted saddle horses and mules for the ascent. Others chose to walk. Most, including Pickering and Mrs. Fleming, rode up in carriages. "Several dangerous turns in the road, at one of which we all had to get out. Road so narrow that there is no chance for teams to pass. Outer wheels within a foot of the edge (and death) for a large part of the way." Those who could bear to look down extolled the view of orange groves and vineyards in the valley.

At the top, Pickering, hoarse from days of talking and the alkali dust of the desert, retreated to the one-room cottage assigned to him. "Living very primitive, but comfortable. No appliances for blacking boots, which are always white with dust, instead of black. A feather duster is used in this part of the country instead of a blacking brush. My greatest wants are a cow and a bath tub. Water is scarce, and milk much more so, as there is no grass on top, and all the fodder must be hauled up the mountain." The recently visited Lowell Observatory, in contrast, had accommodated a dairy cow named Venus.

Most topics of discussion at the Solar Union's plenary sessions pertained specifically to the Sun, naturally enough, in a mix of English, French, and German. Not until the afternoon of the final day, Friday, did the solar scientists vote, unanimously, in favor of extending their studies to the stars and formally considering the question of stellar classification.

"A committee of fourteen is appointed and I am made chairman (they kindly say 'of course'). I rise to express my thanks, and ask the members of the committee to remain after we adjourned so that we might begin work at once. A shout of laughter goes up as all have heard of our meeting with the thermometer at 100°."

Undaunted, everyone present who had been named to the new committee stayed put as Pickering requested, and listened to him tell the story of the Henry Draper classification. He described how the letters of its alphabet had strayed from the usual order into Miss Cannon's arrangement, in which each category seemed to define a different stage in the life of a star. Pickering did not press for the system's acceptance. He anticipated many more discussions before the committee, let alone the Solar Union as a whole, reached consensus on classification. For now he merely wished to acquaint his fellows with the system he knew best, and hear their ideas on how to proceed.

The first to speak, Mount Wilson deputy director Walter Sydney Adams, testified loudly in favor of the Draper system. The ensuing discussion soon proved that most members shared his good opinion of it. "As much to my surprise as to that of the others," Pickering marveled in his diary, "practically everyone approved our system, so that instead of an attempt to replace it, it received the strongest endorsement I could have desired."

CHAPTER NINE

Miss Leavitt's Relationship

THE EASTBOUND TRAIN TO BOSTON, lacking any reserved cars for astronomers, afforded Pickering few opportunities for politicking. Nevertheless he managed to meet briefly between San Francisco and Denver with two members of his new Committee on the Classification of Stellar Spectra. Together they constructed a questionnaire for polling their peers about the pros and cons of the Draper system. Although the full committee favored the Draper classification, some wanted to modify it, a little or a lot, before proposing its formal adoption at the next meeting of the Solar Union, three years hence, in Bonn.

The beauty of the Draper nomenclature lay in the richness of its data. Harvard's Draper Memorial catalogues comprised more than thirty thousand stars, a claim that no other classification could make. The great number of stars falling into the relatively small number of categories affirmed the system's validity. Its level of complexity struck a pleasing compromise between the minimalism of Secchi and the minutiae of Miss Maury. Moreover, it depended entirely upon observable differences, without defending a particular theory.

Not theorizing had been a point of honor with Pickering from the outset. By 1910, however, young astrophysicists chafed to embrace theory. The ideal classification system must be rigid enough to guide and support new research, yet fluid enough to contain conflicting ideas about the dynamics, distribution, and evolution of stars.

In November committee secretary Frank Schlesinger of the Allegheny Observatory in Pennsylvania sent out the questionnaire he had helped draft on the train. It went to all fifteen committee members and about as many nonmembers selected for their strong interest or expertise in classification—notably Annie Cannon, Williamina Fleming, Antonia Maury, and Ejnar Hertzsprung, the Danish astronomer who had so emphatically endorsed Miss Maury's approach.

The questionnaire began with a recapitulation of the committee's impromptu meeting on Mount Wilson at the close of the Solar Union conference. Given the fact that all present had smiled on the Draper classification as the most useful ever proposed, the first question asked, "Do you concur in this opinion? If not, what system do you prefer?"

Answers trickling in over the next several months overwhelmingly favored the Draper system, as might have been predicted. Even Hertzsprung endorsed it, though he called for specific improvements in response to question number two: "In any case, what objections to the Draper Classification have come to your notice and what modifications do you suggest?"

Here a few astronomers took aim at the system's alphabetical names. Mundane labels such as *B* and *A,* they felt, failed to conjure any helpful images. In contrast, the system fashioned by Norman Lockyer in 1899 applied the name of a typical star in each category to the category as a whole. Procyon, for example, a yellowish star in the Little Dog constellation, defined Lockyer's Procyonian division—a cumbrous but evocative term.

Neither Pickering nor Mrs. Fleming had viewed the alphabet letters as permanent fixtures when they introduced them, but rather as neutral symbols, easily replaced with meaningful names once meanings emerged. Even so, years of use had imbued the letters with significance. At Harvard, at least, the mention of *A* instantly called to mind an alpha star like Altair in the Eagle constellation, with blue-white light and a spectrum of unadulterated hydrogen lines.

Among those astronomers content with capital letters, a few regretted the lack of alphabetical order in the Draper system. They thought the progression O, B, A, F, G, K, M looked grotesque or random, as though signifying nothing. Miss Maury roundly rejected it—not simply on aesthetic grounds, but because she had convinced herself the categories, as now arranged, represented the true course of stellar evolution. The "overwhelming predominance" of types O and B in the nebulous regions of Orion and the Pleiades, she told the committee, proved that stars were born from gaseous nebulae in a blue-white heat. As stars aged, they cooled, faded to white, then yellow, finally ending their days in red senescence. The letters or numbers affixed to each stage, therefore, should reflect the seamless flow of stellar life.

Astronomers who shared Miss Maury's evolutionary view commonly spoke of "early stars" when they meant white ones, and called the red stars "late." Those opposed clung to the color words and cautioned against hitching classification to an evolutionary theory. Henry Norris Russell of Princeton, the youngest member of Pickering's committee, envisioned a different evolutionary path from the one Miss Maury described. Russell thought stars might start off red, warm up to yellow or white, and then cool down to red again. He further theorized that stars led different lives depending on their birth weight, and that only the most massive ever achieved the highest temperatures.

"The Draper Classification seems to me all the better because the letters are not in alphabetical order," Russell declared. "This helps to keep the novice from thinking that it is based on some theory of evolution." Apparently the alphabet could flout its own order and still remain effective—or even improve its utility—as a labeling scheme. Pickering could see that much on his typewriter's QWERTY keyboard.

The third of the questionnaire's five questions contained three parts: "Do you think it would be wise for this committee to recommend at this time or in the near future any system of classification for universal adoption? If not, what additional observations or other work do you deem necessary before such recommendations should be made? Would you be willing to take part in this work?"

The mixed reactions to this question crossed party lines. Some of the most outspoken boosters of the Draper system hesitated to press for its formal adoption, fearing the time not yet ripe. The Draper classification surely surpassed all competition, but perhaps something grander might yet be created in its place.

Committee member Edwin Frost of the Yerkes Observatory had long dreamt of a classification system modeled on the ones for plants and animals, dividing the kingdom of the heavens into phyla, classes, genera, and species, all with Latin names. He still hoped astronomers would set that sort of system as a future goal. For the time being, however, Frost thought it foolhardy to tamper with the Draper classification, especially given Pickering's personality. "With his habitual courteous consideration for the views of others," Frost warned in his questionnaire response, "Director Pickering might adopt those of the suggestions made with some unanimity, and then we should have still another classification to add to the present confusion."

Question four concerned a single detail: "Do you think it desirable to include in the classification some symbol that would indicate the width of the lines, as was done by Miss Maury in *Annals of the Harvard College Observatory,* Volume 28?" This question, too, elicited oddly divided opinions. Both Miss Cannon and Mrs. Fleming gave a qualified yes. Miss Cannon pointed out that such distinctions applied to only a small fraction of stars studied. Mrs. Fleming welcomed any symbol that would obviate the need for extensive remarks.

The open-ended final question asked, "What other criteria for classification would you suggest?" Answers ranged widely, but the most common answer was no answer.

When Pickering apprised Mrs. Draper of the growing acclaim for her husband's namesake system, she declared the case "a triumph"—the well-deserved response to the years of labor the director had devoted to the classification and the thought he had expended upon it. She was happy for him, she said, and happy, too, for the sake of Henry's memory.

MISS CANNON FELT TOUCHED "very vitally" by the adoption issue. "It is a far cry from our own sun to the nearest star," she wrote in a memoir, "yet we know that the stars are suns and that many of them are in exactly the same state of composition as our sun. It is therefore befitting that the Solar Union should be interested in the composition of the heavenly bodies." Though she welcomed their interest, she worried "lest this great international body might adopt one of the several other systems of classification proposed, and not adopt ours."

Mrs. Fleming had founded the first Draper catalogue on thousands of tiny spectra photographed through a prism at the telescope's objective end. Those pictures portrayed the violet end of the spectrum well enough, but captured very little of the red. Now that newer photographic techniques and improved dry plates could cover a wider spectral range, Miss Cannon tested the soundness and durability of the Draper classification by reexamining some of the old stars in new photographs. She took pains to work "blind," first classifying the wide new spectra, and only later looking up Mrs. Fleming's designations. It soothed and gratified her to see the overall agreement between them. Apparently the violet end of the spectrum sufficed to settle a star's identity. Miss Cannon corrected some of the original classifications, but more often she just enhanced them with the added spectral detail at her disposal, such as changing an F star to an F 5 G.

Mrs. Fleming helped revise the ever-improving Draper Catalogue by revisiting the numerous spectra formerly lumped in the "peculiar" category. The pace of her variable star discoveries remained slow as she hastened the delivery of *Annals* volumes to the printer. That winter she found "only" eight. Early in 1911, however, in recognition of her cumulative record, the Sociedad Astronómica de México awarded her its Guadalupe Almendaro gold medal for her prowess among the variables. The Bruce Medal still eluded her, but she hardly lacked for recognition from her fellow members of the Astronomical and Astrophysical Society of America, or the fans who had made her an honorary member of the Royal Astronomical Society and the Société Astronomique de France.

Mrs. Fleming so often visited Professor Sarah Whiting's classes at Wellesley as a guest lecturer that the college had named her its Honorary Fellow in Astronomy. She was anticipating giving another talk at Wellesley, scheduled for late May, when the fatigue that had dogged her all spring turned to malaise. She elected to enter a hospital for rest, but, once there, her condition worsened, and she developed a fatal pneumonia. Edward Fleming, now the chief metallurgist for a large copper company in Chile, could not get to Boston in time to see his mother before she died on May 21, 1911. She was fifty-four years old and had devoted thirty of her years to the observatory.

"I have just this moment, to my very deep regret, seen in 'Science' a notice of Mrs. Fleming's death," Henry Norris Russell wrote to Pickering from Princeton on June 2. "I cannot do less than write at once to express my sympathy for the loss that I know will be felt very deeply and in many ways by all the circle at the Harvard Observatory, and by a far wider one of friends outside." The youthful Russell had spent time with Mrs. Fleming at meetings in Cambridge and on the previous summer's trek to Pasadena for the Solar Union. "Her loss will be a severe one to science, and must be a terrible blow to her friends," he commiserated. "I did not know the son of whom she spoke so often, and so can hardly send a message to him; but the sense of loss on hearing that she is gone was so keen that it seemed natural to write to you."

Pickering's eulogy for Mrs. Fleming in the *Harvard Graduates' Magazine* retold parts of the saga she had shared with him over the years about her ancestors, the Claverhouse "fighting Grahams"—how her great-grandmother had eloped with Captain Walker of the 79th Highlanders, followed him to Spain in the Peninsular War, then given birth to a son on the Corunna battlefield the very day the captain was killed in combat. Surely the family mettle had stiffened Mrs. Fleming's spine. As her long-time supervisor, Pickering could aver, "It was only necessary to tell her exactly what was needed, and she saw that it was carried through successfully in every detail." After enumerating her many astronomical discoveries and distinctions, he said she "formed a striking example of a woman who attained success in the higher paths of science without in any way losing the gifts and charm so characteristic of her sex."

Miss Cannon had written the obituary notice for Mrs. Fleming that Henry Norris Russell read in *Science,* as well as another, longer one for the

Astrophysical Journal. The articles gave her occasion to praise her late friend's "naturally clear and brilliant mind," "her extremely magnetic personality," and "that quality of human sympathy which is sometimes lacking among women engaged in scientific pursuits." Miss Cannon also took pains to describe the rare collection of glass plates entrusted to Mrs. Fleming's reliable care: "Each photographic plate may be likened to the only existing copy of a valuable book, and, being very fragile, must be safely stored, and at the same time must be accessible, so as to be consulted readily at any moment."

It seemed right that Miss Cannon should succeed Mrs. Fleming as the official curator of astronomical photographs. Pickering raised the idea in October 1911 with Harvard's new president, Abbott Lawrence Lowell (brother of Percival Lowell), who took office after Charles Eliot retired in 1909. Not only had Miss Cannon fulfilled the duties of curator since Mrs. Fleming's passing, Pickering said, but she had done so "in a very satisfactory manner." Moreover, he added by way of endorsement, "Miss Cannon is the leading authority on the classification of stellar spectra, and perhaps on variable stars."

Lowell reacted negatively. "I always felt that Mrs. Fleming's position was somewhat anomalous," he replied on October 11, "and that it would be better not to make a regular practice of treating her successors in the same way." He therefore declined to recommend that the Harvard Corporation appoint Miss Cannon. Instead, he suggested, Pickering should install her himself as a matter of ordinary department business, with less fuss, lower pay, and no cause to cite her name in the university catalogue.

The members of the Visiting Committee were appalled. "It is an anomaly," their 1911 report said of the slight to Miss Cannon, "that, though she is recognized the world over as the greatest living expert in this line of work, and her services to the Observatory are so important, yet she holds no official position in the University."

Miss Cannon did not let the denial of university title stand in the way of duty. In October 1911 she embarked on new projects to unify and strengthen the Draper system. She reclassified Miss Maury's bright northern stars, to conform their Roman numerals to the current Draper designations. She took up Mrs. Fleming's unfinished last catalogue of faint southern stars and brought its 1,688 listings into alignment. Both her speed and her certainty in judgment had increased, as, too, had her love for the work. She thought she might as well keep going, examine more plates, continue classifying indefinitely, expand the Draper Catalogue exponentially.

PICKERING'S VOLUNTEER ARMY of variable star observers blanketed the Northeast by 1911, and extended as far west as California. There was even an outlier in Australia. Faculty and students at New England colleges such as Amherst, Vassar, and Mount Holyoke participated energetically in the routine observations. Strong foreign support arrived monthly from amateurs in the Variable Star Section of the British Astronomical Association. Harvard's own professional staff still led the charge, with Leon Campbell alone averaging a thousand observations per month through a 24-inch reflector.

Campbell's attentions shifted in the spring of 1911, when Pickering sent him to Arequipa as director of the Boyden Station. The new post positioned Campbell to keep vigil over the long-neglected long-period variables of the southern sky, but also forced his abandonment of the northern ones. To fill Campbell's vacancy, Pickering called on the corps of volunteers. He drew up a list of 374 variables requiring frequent surveillance, and assigned each star to one or more regular observers. He also circulated the list as an invitation for others to participate. Given the interruptions to be expected from inclement weather, moonlight, and personal engagements, one star could never have too many pairs of eyes on it. He prepared printed forms to facilitate the filing of reports, provided finder charts to help new recruits locate their stars, and promised to publish the volunteers' observations. Hoping to head off any needless duplication of effort, Pickering urged his troops to communicate among themselves and cooperate wherever possible, such as by observing at different times of the month and different hours of the night.

Anna Palmer Draper funded the Harvard project to photograph the spectra of the stars—the unfulfilled dream of her late husband. She sat for this portrait by John White Alexander in 1888.

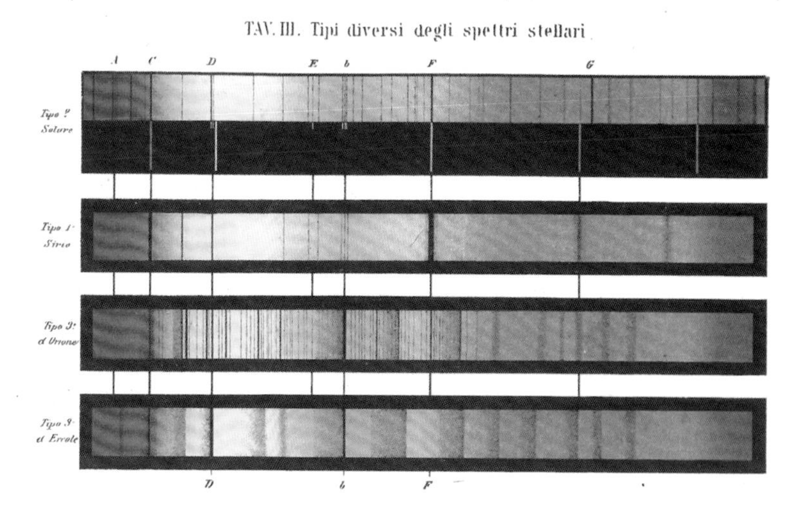

Dark Fraunhofer lines cutting across the rainbow spectra of the Sun and other stars gave Father Angelo Secchi of the Vatican Observatory a means of categorizing the various stellar types. This image from his 1877 book, *Le Stelle: Saggio di Astronomia Siderale,* shows examples of the classes he identified.

The expedition party that gathered in Rawlins, Wyoming Territory, to view the total solar eclipse of July 29, 1878, included (from far right) English astronomer Norman Lockyer, Thomas Edison, and Henry and Anna Draper.

The dome of the Great Refractor dominated the appearance of the Harvard College Observatory in the 1870s. A smaller telescope was mounted in the west wing.

Edward Charles Pickering became the director of the observatory in 1877 and served for more than forty years as its visionary leader.

Williamina Paton Stevens Fleming began working for the Pickerings as a maid, but later went on to establish a system for classifying stars by their spectra.

Mrs. Fleming (standing at rear) earned a supervisory role over the other female computers and also a coveted Harvard title as curator of astronomical photographs.

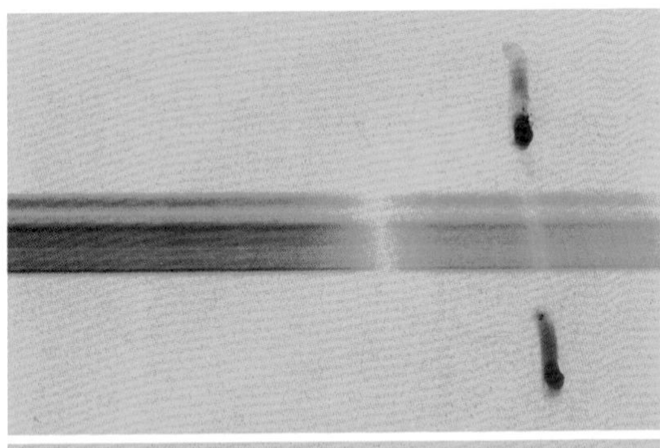

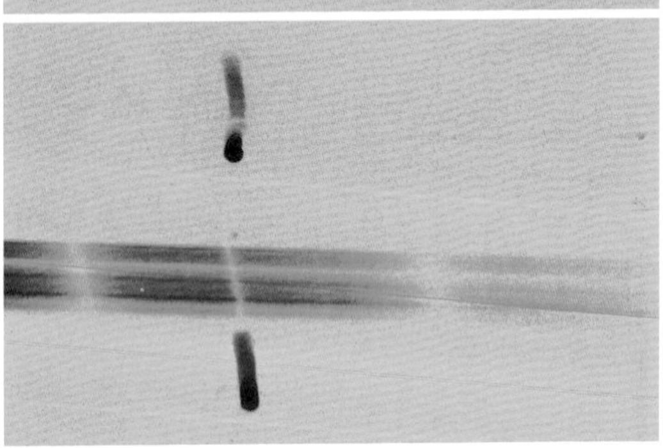

Blue ink squiggles on these paired spectra of the star Mizar reveal a doubled K line in the top image and a single K line in the bottom one—differences that led Edward Pickering to his 1887 discovery of the first spectroscopic binary.

In order to photograph the stars of the Southern Hemisphere, Harvard established an auxiliary observatory, the Boyden Station, in Arequipa, Peru, within sight of the dormant volcano El Misti. The observer's house built by William Pickering is at right.

Annie Jump Cannon, a Wellesley College graduate, was continuing her astronomy studies at Radcliffe and also assisting at the Harvard Observatory when this photograph was taken, ca. 1895.

Antonia Maury (far right) and her sister Carlotta (far left) are pictured here with their Aunt Ann (Mrs. Daniel Draper) and young cousins Harriet and Dorothy Catherine. The woman in the dark dress is unidentified.

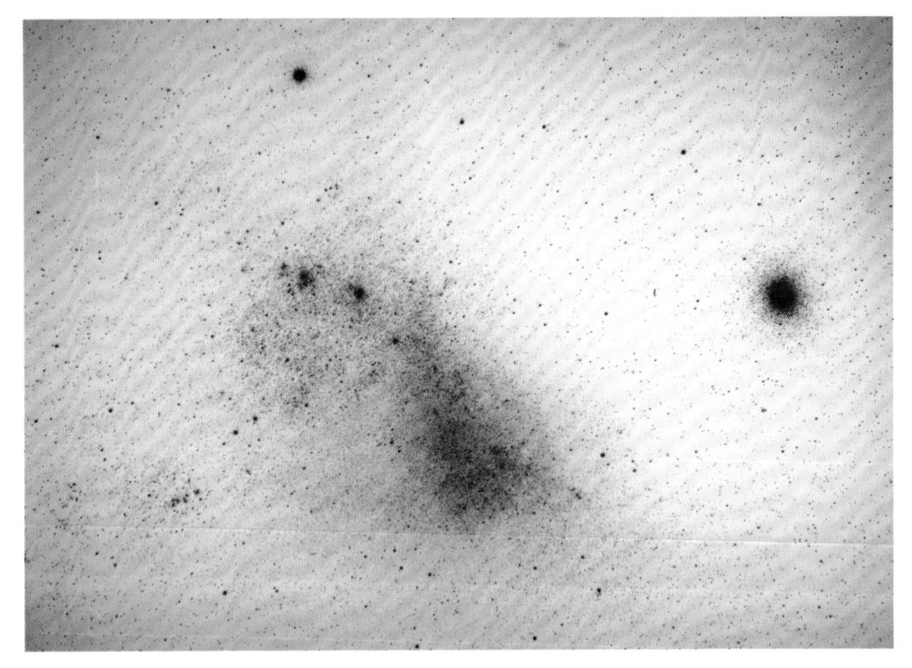

Stars appear as black dots in this negative plate of the Small Magellanic Cloud, a satellite galaxy of the Milky Way that can be seen from the Southern Hemisphere. The splotch at right is the dense globular cluster of stars known as 47 Tucanae.

Called a fly spanker for its resemblance to a fly swatter, this diminutive instrument helped computers compare the relative brightness of stars.

Henrietta Swan Leavitt discovered a relation between certain stars' peak brightness and the time it took them to cycle through their changes in magnitude. This "period-luminosity relation," also called the Leavitt law, provides a means for measuring distances across space.

Many noted foreign astronomers joined this Harvard gathering in August 1910. Miss Cannon, in a white dress, is at far left. Leon Campbell kneels just in front of Miss Cannon, and sitting in front of him is Winslow Upton, librettist of *The Observatory Pinafore*. The woman next to Miss Cannon is Lucy May Russell, wife of Henry Norris Russell, who is at her side. Pickering stands front and center, Mrs. Fleming, wearing all black, also in the front row, and Henrietta Leavitt just behind her, in white. Solon Bailey, bald and bearded, sits at far right.

Pickering poses at the entrance to the Brick Building with the female staff, ca. 1911. Margaret Harwood is at far left, Arville Walker just in front of her. Ida Woods stands at far right in the front row. The white-haired lady on the step behind her is Florence Cushman. At her right is Annie Cannon, and Evelyn Leland is in the back row between them.

This 1918 chain of Harvard assistants begins with Ida Woods (at far left) and links to Evelyn Leland, Florence Cushman, Grace Brooks, Mary Vann, Henrietta Leavitt, Mollie O'Reilly, Mabel Gill, Alta

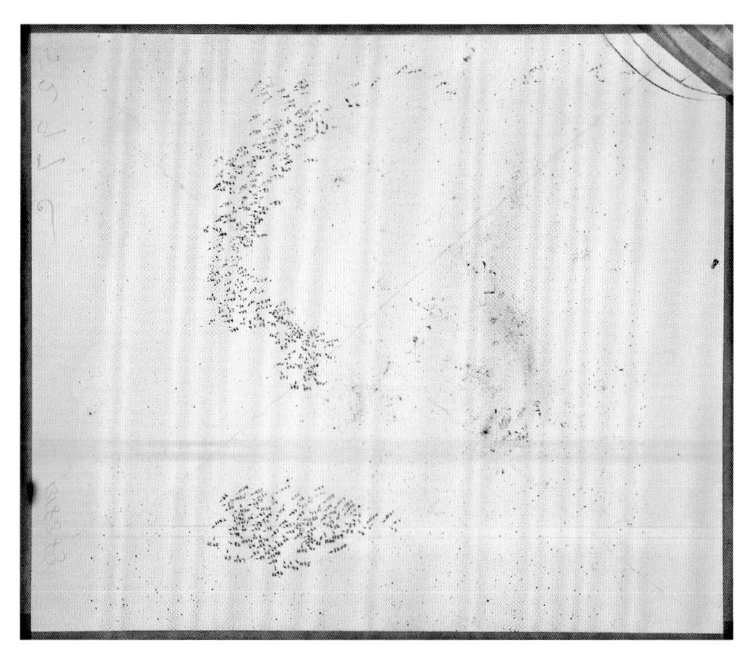

This two-hour-long exposure of the Large Magellanic Cloud, taken with the 8-inch Bache telescope at Arequipa on January 23, 1897, gave the staff in Cambridge hundreds of objects to number and ponder over a period of years.

Carpenter, Annie Cannon, Dorothy Block, Arville Walker, and telescope operator Frank E. Hinkley, ending (at far right) with chief of stellar photography Edward King.

The Brick Building, where the glass plates were stored, became a second home to Annie Cannon (left), who classified more than a quarter of a million stars by their spectra, and her colleague Henrietta Leavitt, who sought out variable stars and monitored their behavior.

In a typical work session, Miss Cannon would write numbers next to all the spectra on a plate, then call out each number and also her judgment of its spectral type to a recorder who wrote down her pronouncements.

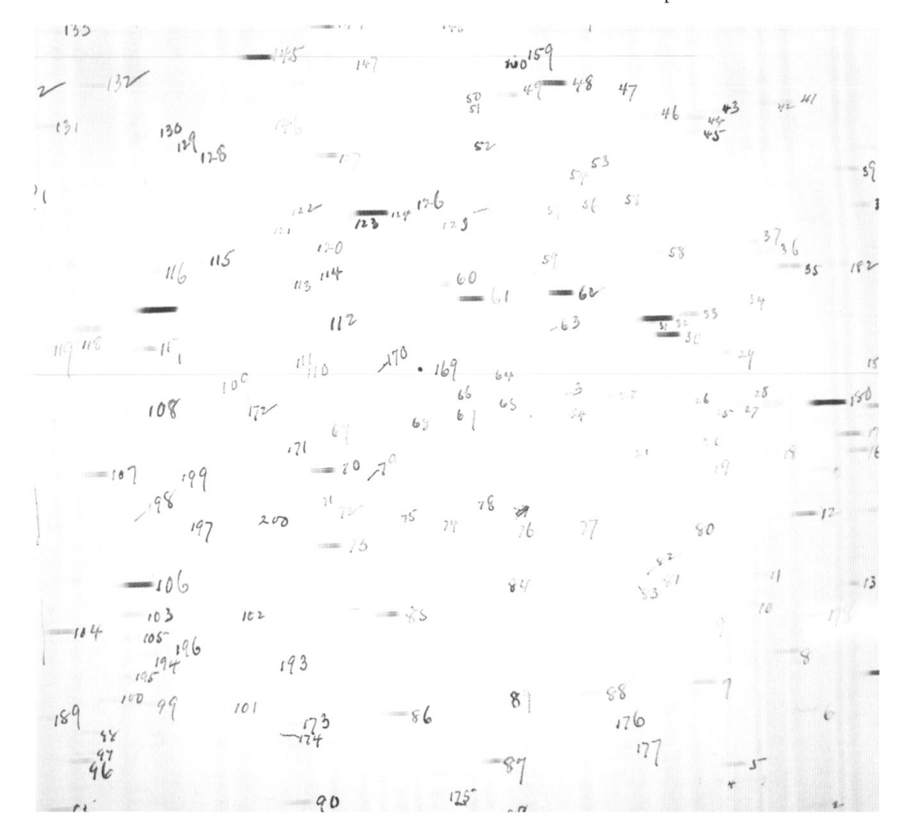

Soon after Harlow Shapley took over as director of the observatory, Annie Cannon accompanied Solon and Ruth Bailey to Peru, where she often walked (or rode) for hours during the day and then observed late into the night.

Miss Cannon said she did not mind climbing up and down ladders to operate the 13-inch Boyden telescope and take her own plates of the southern stars.

Harlow Shapley enjoyed the flexibility of the unique revolving desk-and-bookcase combination devised by his predecessor, Edward Pickering.

Cecilia Payne traveled to the Harvard Observatory from Cambridge University in England, where she had been inspired by Arthur Stanley Eddington to devote herself to astronomy.

Miss Payne (right) and Adelaide Ames (center), called the "Heavenly twins," welcomed Harvia Hastings Wilson as the third graduate student in 1924. Miss Payne went on to earn a Ph.D.—the first doctoral degree in astronomy awarded at Harvard.

Margaret Harwood sat on the floor for this posed tableau taken on May 19, 1925. Harvia Wilson is at far left, sharing a table with Annie Cannon (too busy to look up) and Antonia Maury (left foreground). The woman at the drafting table is Cecilia Payne.

The 1929 New Year's Eve performance of *The Observatory Pinafore* featured (from left to right) Peter Millman; Cecilia Payne as Josephine; Henrietta Swope, Mildred Shapley, Helen Sawyer, Sylvia Mussells, and Adelaide Ames as the computer chorus; and Leon Campbell in the role of Professor Searle.

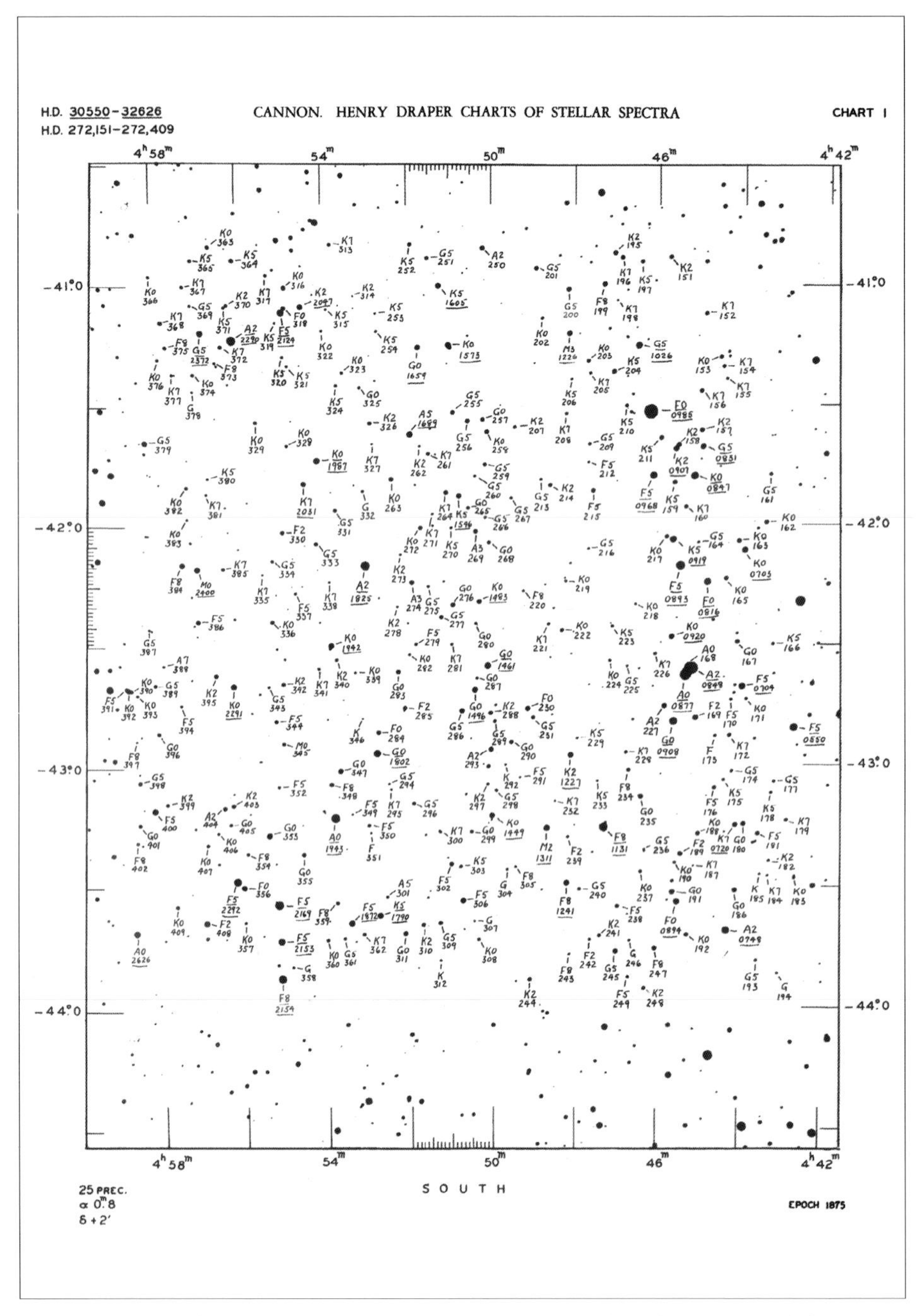

The switch from tabulated columns of numbers to the chart format lowered the cost of publishing Miss Cannon's classifications for the Henry Draper Extension.

Antonia Maury installed a 6-inch Clark telescope at the old Draper homestead in Hastings-on-Hudson. She intended it for the edification of local residents, especially children.

As youngsters, Katherine and Edward Gaposchkin played in and around the observatory, within easy reach of their parents, Sergei Gaposchkin and Cecilia Payne-Gaposchkin.

Roughly half a million glass plates tilt to the left and right on shelves inside metal cabinets at the Harvard Plate Stacks. Each plate's paper jacket identifies the date the photograph was taken, the celestial area covered, the telescope employed, the duration of exposure, the condition of the sky, and other pertinent information.

The Great Refractor stands idle today inside its large dome. Two people can still sit side by side in the commodious, adjustable observing chair designed by founding director William Cranch Bond, but the skies over Cambridge are no longer dark enough at night to permit new discoveries.

Popular Astronomy editor Herbert C. Wilson saw the need for an even higher order of organization among variable stargazers. In the August–September 1911 issue of the magazine, Wilson entreated his readers, "Can we not have in America an association of observers with a 'Variable Star Section,' a 'Jupiter Section,' etc.?" In almost instant reply, lawyer and avid amateur observer William Tyler Olcott of Norwich, Connecticut, announced the October formation of the American Association of Variable Star Observers (AAVSO).

Olcott had caught the variable star fever from Pickering, at a public lecture the director gave in 1909. The two corresponded afterward, and Pickering, recognizing Olcott's dedication, arranged for Leon Campbell to coach him at his Connecticut home. The founding of the AAVSO cemented the already close ties between Olcott and Harvard.

Professor Anne Sewell Young of Mount Holyoke, one of Pickering's most reliable regulars, immediately signed up as a charter member of Olcott's association. In December 1911 her recent observations formed part of the AAVSO's first published report in the pages of *Popular Astronomy.* Soon Sarah Frances Whiting and her assistant Leah Allen of the Wellesley College observatory joined the AAVSO, and also Maria Mitchell's successor at Vassar, Caroline Furness. The group welcomed devotees from any sort of day job. Charles Y. McAteer, for example, worked as a locomotive engineer for the Pittsburgh, Cincinnati, Chicago and St. Louis Railway Company. At the end of the night freight run into Pittsburgh, he would go home to the 3-inch telescope in his backyard and observe variables till dawn.

The AAVSO members concentrated on the variables of long period. Most such stars waxed and waned gradually through as many as nine magnitudes of change in the course of a few months to a year-plus. They were always on their way up or down the brightness scale, filling hours with quiet purpose for Pickering's minions. The short-period variables, on the other hand, defied tracking by telescope. Within days—within hours in some cases—they suddenly flashed, then faded. The rest of the time they stayed quiescent at their lowest ranges. One needed great good luck or a series of photographic snapshots to glimpse their brief brightening. Just such a series of pictures taken within two to three days of each other in 1905 had alerted Miss Leavitt to the untold number of quick-change stars in the Magellanic Clouds.

MISS LEAVITT HAD BEEN CALLED HOME AGAIN to Wisconsin after her father died on March 4, 1911, and she spent the spring and summer helping her mother settle Reverend Leavitt's small estate. Back in Cambridge in the fall, she found the observatory family still adjusting to the loss of Mrs. Fleming. Miss Cannon was supervising the computers. Mabel Gill, a staff member since 1892, had taken over the preparation of several *Annals* volumes for the printer, and, together with another experienced coworker, Sarah Breslin, was wrapping up Mrs. Fleming's longstanding effort to measure the variables she had discovered against the 222 star sequences she tailor-made for that purpose. Miss Maury had fled once more to the old Draper homestead at Hastings-on-Hudson.

Resuming the hunt for new variable stars on the Harvard sky maps, Miss Leavitt continued to ponder the thousands she had encountered in the Magellanic Clouds.

The prevalence of variables in those two southern star Clouds beggared all comparison. Miss Leavitt had tallied more than nine hundred in the Small Cloud and eight hundred in the Large, without even venturing into the Clouds' centers, where stars clumped inseparably together.

"If the stars were equally dense over the whole sky," Solon Bailey guessed, "their number would exceed ten billions, and the sky would be so luminous that there would be no real night." Bailey had scanned the southern sky at Arequipa, from the decks of ships in the south latitudes, and on the Great Karroo of South Africa. In the perfect darkness of those remote sites, he had seen the star-spangled Milky Way spill across the night's horizons. His telescopes had collapsed the vast distances to immerse him in that river of stars. Miss Leavitt, denied such intimacy with the heavens, could only imagine herself standing agape in the Andes, under the southern meanders of the Milky Way, watching the Magellanic Clouds trail after the star stream like a pair of lost sheep.

Bailey believed the two Clouds to be unique structures, separate from the Milky Way. If so, if they in fact existed outside the bounds of the galaxy, then each Cloud constituted its own so-called island universe. Possibly the numerous other splotchy white nebulous objects scattered through space were also separate star systems, independent of the Milky Way.

Bailey's two- and four-hour exposures of the Magellanic Clouds, taken with the Bruce telescope, had revealed crowds of stars as faint as seventeenth magnitude. Miss Leavitt picked her way among them in her initial study by repeating the Baileys' globular-cluster strategy: She ruled a reticule of one-centimeter squares on a glass plate, rendering it a transparent sheet of graph paper. Then, superposing the reticule on images of the Clouds, she cordoned off small groups of stars and charted them through an eyepiece fitted with micrometrical crosshairs.

Immune to all distractions, she differentiated the individual members, numbered them, recorded their relative positions, and tracked the variables' brightness changes through time. The proximity of variables to one another complicated her task, as did their distances from suitable comparison stars. The pattern of the variables' alteration also challenged her, since most of them remained at their dimmest most of the time, brightening suddenly in short bursts. In her 1908 publication, "1777 Variables in the Magellanic Clouds," she gauged all the ranges of magnitude and gave the maximum and minimum value for every star, as best she could. She traced the complete light curves of only sixteen stars, yet this small, select sample (one one-hundredth of the whole) had shown an intriguing trend: the brighter variables had the longer periods, as though the one thing depended on the other.

Since the sixteen variables all belonged to the compact Small Cloud, Miss Leavitt reasoned they all lay roughly the same distance from Earth, just as all her relatives in Beloit lived about equally far from Cambridge. Therefore the ones that *looked* brighter must actually *be* brighter.

The unexpected correlation of brightness with period could be mere coincidence, as Miss Leavitt well knew. But if the same pattern held true for a larger number of similar variables, then the correlation itself might indicate something stupendous.

In 1911 Miss Leavitt tracked the step-by-step variation in another nine stars on the glass plates. As before, the brighter variables took the longest times to cycle through their variations. She plotted the numbers on a graph, with period lengths along the *x* axis, maximum and minimum magnitudes on the *y*. Connecting the dots, she got two smooth curves, and when she reduced the periods to a logarithmic scale, her curves snapped into straight lines. The trend among Miss Leavitt's stars was real. Pickering called it "remarkable" when he announced her results in a *Harvard College Observatory Circular* on March 3, 1912. He used the word "law" to describe the finding she had demonstrated for twenty-five stars in the Small Magellanic Cloud: The brighter the magnitude, the longer the period. It meant that certain types of variables telegraphed their true magnitudes in the duration of their light cycles. Those stars heralded the coming of distance markers in the farther reaches of space. As soon as astronomers learned the key to the stellar code—the degree of brightness linked to each period—they could determine the stellar magnitudes by watching a clock, then leap the interstellar distances on the wings of Isaac Newton's inverse square law: a variable only one-quarter as bright as another of the same period must lie twice as far away.

Ejnar Hertzsprung in Denmark seized on Miss Leavitt's period-luminosity relation. He, too, had been drawing graphs, plotting one stellar characteristic against another to test their interdependence. Like many but not all of his contemporaries, Hertzsprung saw the Draper spectral classification as a temperature gradient: The blue-white O stars were the hottest, the red M the coolest. Two red stars of nearly identical spectra therefore shared the same temperature; if one of them looked brighter than the other, then it must be either closer or larger. Hertzsprung could often judge relative distances between two such stars by their proper motion. If the farther one—the star that moved less—was the brighter of the two, then it necessarily had more surface area from which to radiate its light. This reasoning had opened Hertzsprung's mind to the possibility of exceptionally large stars, or giants. In the past he had lauded Miss Maury for noticing the spectral nuance that could separate giants from dwarfs. Now he thanked Miss Leavitt for the means to measure distances previously beyond reach.

Hertzsprung identified about a dozen examples of stars like Miss Leavitt's within the Milky Way. They followed the same kind of light curve, with a steep rise to peak brightness giving way to gradual decline. These stars shone many orders of magnitude brighter than their peers of the same period among Miss Leavitt's stars. The differences put the Small Magellanic Cloud, by Hertzsprung's reckoning, at a distance of thirty thousand light-years—a chasm so great as to strain credulity.

Henry Norris Russell followed some of Hertzsprung's same ideas to similar conclusions about size, brightness, and distance. Based on his own calculations, Russell posited that Miss Leavitt's variables and their yellow counterparts in the Milky Way were all giants.

Miss Leavitt herself did not pursue these lines of research. She advanced the hunt for new variables in her third of the sky and went on fine-tuning the magnitudes of the North Polar Sequence, allowing others to build on the strength of her relationship.

On Nantucket Island off the coast of Massachusetts, where Maria Mitchell had made her world-renowned comet discovery in 1847, a small observatory commemorated her good name. The Maria Mitchell Association occupied the astronomer's birthplace on Vestal Street and a domed structure next door to it. The association was founded in 1902, thirteen years after Miss Mitchell's death, by her cousin Lydia Swain Mitchell, who had also been born in the house on Vestal Street. Cousin Lydia, now Mrs. Charles Hinchman, lived in Philadelphia with her husband and children, but returned to Nantucket every summer, and felt a duty to keep the spirit of astronomy alive on the island. She frequently consulted the director of the Harvard College Observatory for advice, and asked him to recommend guest lecturers. For several years beginning in 1906, Annie Jump Cannon made a summer pilgrimage to Nantucket in this capacity. She also taught a correspondence course in astronomy and helped bring Ida Whiteside of the Wellesley Observatory and Professor Florence Harpham of the College for Women at Columbia, South Carolina, to the island as "summer observers." At "Moon Nights," the public got to look up through Maria Mitchell's own 3-inch comet-seeker and also the 5-inch Alvan Clark telescope purchased for her in 1859 by a group of admirers called the Women of America. The popularity of the Vestal Street activities gave Mrs. Hinchman the idea of awarding a yearlong stipend to a young lady who could conduct research at the observatory while instructing the locals in star lore. An appeal to Andrew Carnegie garnered $10,000 toward the goal. In 1912 Margaret Harwood, a Harvard computer, received the first $1,000 astronomical fellowship of the Nantucket Maria Mitchell Association.

Miss Harwood had joined the Harvard staff in 1907, recruited from that year's Radcliffe graduating class by her astronomy professor, Arthur Searle. Having boarded with Arthur and Emma Searle and their daughters since her freshman year of college, Miss Harwood was a familiar face at the observatory even before day one of her employment. At first she assisted Searle, whom she called her "father in astronomy," by computing the orbits of comets. She helped Miss Leavitt assess the photographic magnitudes of circumpolar variables on the glass plates, and learned from Miss Cannon how to observe them by telescope. Pickering enlisted her aid in recalculating the positions of sixteen thousand stars catalogued by the Bonds in the 1850s.

The Maria Mitchell Association invited Miss Harwood to carry on her research at Harvard for the first half of her fellowship year, and then transfer in June to Nantucket, where they accommodated her through December in an upstairs bedroom of the Mitchell residence. Downstairs in the museum proper, her housemates included collections of Nantucket flora and fauna, fossil displays, and a library about equally divided between astronomy and natural history topics. On Monday nights in the parlor she gave a lecture that spilled out onto the lawn or into the next-door observatory for stargazing. When Professor Pickering visited, he declared the site, remote from the smoke and glare of cities, ideally suited to the study of asteroids by photographic telescope, and the association raised the funds to purchase such an instrument. Miss Harwood so thoroughly endeared herself to the Mitchell family members and the Vassar alumnae who endowed her position that they all looked forward to her return for a second season in the summer of 1913.

On another island, far south of Nantucket, William Pickering established a one-man observatory at Mandeville, Jamaica. William had first tested Jamaica as an observing site in 1899, when he repaired there for a family vacation and noticed a pleasant clarity in the air. In October 1900, after convincing his brother of Jamaica's suitability as an outpost in the Eros campaign, William returned for a six-month stay with his family and a new telescope at a rented Mandeville estate named Woodlawn. Unfortunately, he failed to get good pictures of Eros. In an attempt to salvage the expedition, he remained at Woodlawn through August 1901, photographing the Moon for a lunar atlas that he later published.

Over the next several years, while William maintained a base at Cambridge, he pursued the Moon and planets from sites in California, Hawaii, Alaska, the Azores, and the Sandwich Islands. In 1911, when his risky investments failed and ruined William financially, Edward helped him relocate temporarily to the familiar Woodlawn grounds. The 11-inch Draper refractor accompanied William to what he euphemistically termed the Harvard Astronomical Station in Jamaica. Woodlawn, once a thousand-acre plantation, owned an ideal telescope mounting site on a patio previously used to dry coffee beans. William's $2,500 annual salary stretched further in the Caribbean than in Cambridge, and he pronounced the seeing at Mandeville the equal of Flagstaff's or Arequipa's. He could see no reason to leave. Thus the tropical Woodlawn estate became the Woodlawn Observatory, and William Pickering its astronomical lord of the manor. Isolated and increasingly eccentric, he spoke out as he pleased about the Martian canals, the green vegetation on the red planet, and the likelihood that Mars supported some kind of animal life.

MISS CANNON HAD CLASSIFIED one hundred thousand stars when she set the work aside to spend the summer of 1913 in Europe with her sister, Mrs. Marshall. They planned to attend three major astronomy meetings on the continent, plus all the banquets, garden parties, excursions, and entertainments that such international congresses entailed. On her previous trip to Europe, with her friend and Wellesley classmate Sarah Potter in 1892, Miss Cannon had made the grand tour of popular tourist destinations, camera in hand. This time she would go as a respected astronomer and the only female officer in her professional organization. At the 1912 meeting of the Astronomical and Astrophysical Society of America, the members had voted to change their name to the American Astronomical Society and to make Miss Cannon their treasurer. Now she would seek out her foreign colleagues, many of whom she knew only by reputation or correspondence, in their native settings.

"There are no women assistants," Miss Cannon noted of the Royal Observatory, Greenwich. Travel broadened her appreciation for the singularity of Harvard's large female staff, although she easily befriended men wherever she went. At Greenwich, "Without the slightest feeling of being out of place, without the smallest tinge of embarrassment, I discussed absorbing work with one and another." That evening the astronomer royal, Frank Dyson, called for Miss Cannon and Mrs. Marshall at their London hotel and escorted them to a soiree at Burlington House, the headquarters of the Royal Astronomical Society and four other scientific fraternities. "Never has it been my good fortune to have such a kindly greeting, such hearty good will, such wonderful feeling of equality in the great world of research as among these great Englishmen." At the society's meeting a few days later, she gave a formal presentation about her recent investigation into the spectra of gaseous nebulae.

Mrs. Marshall understandably avoided the scientific sessions, at which Miss Cannon inured herself to being the sole woman in a roomful of as many as ninety men. In Germany, she reported, "Not a single German woman attended these Hamburg meetings" of the Astronomische Gesellschaft. "Once or twice, two or three would come in for a few minutes but I was generally the only woman to sit through a session. This was not so pleasant but at the recesses the men were so kind that nothing seemed to matter, and at the luncheon women appeared in great numbers."

In Bonn, where the Solar Union gathered from July 30 to August 5, the astronomers were treated to a flyby visit of a military zeppelin, a side trip to the Gothic cathedral at Cologne, a riverboat ride up the Rhine, and a gala night in the Bonn observatory that prompted the English-speaking delegates to sing "They Are Jolly Good Fellows" to Director Friedrich Küstner and his wife and daughters. "Luncheon and indeed all meals in Germany," observed Canadian astrophysicist John Stanley Plaskett, "are a much more important and solemn function than with us and take at least twice the time."

Pickering, an elder statesman in this community, spoke at several banquets during the week. He shared impressions of his previous stays in Bonn, a city he had long regarded as the world capital of photometry. It was here that the legendary Friedrich Wilhelm Argelander assembled the Bonner Durchmusterung star catalogue and perfected the Argelander method of studying variables by comparing them to their steady neighbors. Argelander's own small telescope, still mounted at the Bonn observatory, proved an object of veneration for the visiting astronomers.

Only about half the members of Pickering's Committee on Spectral Classification, first convened at Mount Wilson, had come to the Bonn meeting. Those present included Henry Norris Russell, Karl Schwarzschild, Herbert Hall Turner, and of course Küstner, of the local observatory. They met Thursday afternoon, July 31, to polish their report before Friday's discussion and vote. The group had considered incorporating some symbols into the Draper classification that would account for the widths of spectral lines, but ultimately rejected the idea. Rather than retrofit the Draper system, they preferred to look forward and explore the possibility of an entirely new design for stellar taxonomy.

On Friday morning Chairman Pickering read the committee's recommendation to the full assembly at the Physical Institute. He proposed postponing "the permanent and universal adoption" of any system until the committee could formulate a suitable revision. In the interim, however, everyone should foster the well-known and widely praised Draper classification. Approval of the resolution was swift and unanimous. Ditto the subresolution regarding a refinement originally suggested by Ejnar Hertzsprung and already practiced by Miss Cannon. It consisted of a zero subscript for lone letters. Going forward, A_0 would denote a star of purely A-category attributes, showing no B tendencies whatever. The new A_0 reduced plain A to a "rough" categorization.

At the final session on August 5, the Solar Union dissolved its old committees and regrouped into new ones for the work to be done over the next three years, before they would all meet again in Rome.

"When the names of committees were read," wrote Miss Cannon, "I was very much surprised to find that I was put on the Committee on Classification of Stellar Spectra—and one of the novel experiences of the summer was to meet with this Committee. They sat at a long table, these men of many nations, and I was the only woman. Since I have done almost all the world's work in this one branch, it was necessary for me to do most of the talking."

CHAPTER TEN

The Pickering Fellows

YEAR-END HOLIDAY GREETING CARDS from the Harvard College Observatory in 1913 featured a single gold star, with a word at each point naming the five stellar data: position, motion, brightness, spectrum, and color. On Miss Cannon's card, Professor Pickering penned in his personal best wishes "for a Merry Classification and a Happy New Type of Spectrum." Miss Cannon was classifying or reclassifying approximately five thousand stars per month. Along the way she reinstated two of Mrs. Fleming's categories, N and R, putting R ahead of N. The alphabetical disorder of her system was ultimately conquered by a new mnemonic: Oh, Be A Fine Girl, Kiss Me Right Now!

Astronomers from all over Europe and the United States, while awaiting publication of the revised Draper catalogue, often queried Miss Cannon about the spectrum of particular stars for their studies. One of her many regular correspondents, Herbert Hall Turner of Oxford, tendered his congratulations on March 13, 1914, regarding an honor "unanimously and cordially" conferred on her that very day. She received no official notice, however, until early May, when an embarrassed Arthur Stanley Eddington, secretary of the Royal Astronomical Society, apologized to Miss Cannon for the oversight.

"A form of Diploma is being prepared, which you will receive before long," Eddington promised, "but, of course, it was intended that you should be informed at once. The mistake seems to have been due to a misunderstanding between the President [Edmond Herbert Grove-Hills] and myself as to which one of us had undertaken the letter. Our best excuse is that the election of an Honorary Member is so rare an occurrence that we have no routine way of procedure, so that mistakes are not provided against." Eight years had passed since Mrs. Fleming's election, and by 1914 all the previous honorees, save for Lady Margaret Huggins, had passed on.

Miss Leavitt's work likewise drew wide attention, though not the formal accolades of the kind Miss Cannon reaped. Nor did Miss Leavitt travel to international gatherings, but remained at the observatory, sometimes in a supervisory capacity when other officers were away. Bailey, who often played that role, commended her fitness for it, with her nature "full of sunshine" and her ready perception of everything "worthy and lovable in others."

The stars that had led Miss Leavitt to her period-luminosity relation were called Cepheids, for the group archetype, Delta Cephei, in the constellation Cepheus, the King.* John Goodricke of England was the first, in 1785, to describe Delta Cephei's pattern of variation—the sharp rise and slow ebb of brightness that proved characteristic of variable stars in other constellations as well. Some thirty Cepheids were known in the 1890s, before Solon Bailey began discovering scores more in the star clusters of the Southern Hemisphere. Later Miss Leavitt made their number legion. By mid-January 1914, when she had finished counting the variables in her third of the sky and concluded her many years of work on the North Polar Sequence, the Cepheids were beckoning new followers.

Harlow Shapley, a young American astronomer completing graduate studies under Henry Norris Russell at Princeton, visited Harvard in March 1914. Pickering welcomed him in typical fashion, offering to provide any materials from the observatory that he wanted. Miss Cannon took him home to dinner, and when he went to call on Solon Bailey, upstairs by

* In mythology, Cepheus was the father of Andromeda, the chained woman. In the sky, these two lie on either side of Queen Cassiopeia, Andromeda's mother.

the observatory dome, he received advice that shaped the course of his career.

"Bailey was pious and kind, a wonderful sort of man," Shapley later rhapsodized, "but so New England it made you ache." Shapley came from rural Missouri, and had worked as a crime reporter for a Kansas daily before pursuing his higher education. According to Shapley's account of their conversation, possibly recorded in newspaperman's shorthand, Bailey said, "I hoped you wanted to come up here; I have been wanting to ask you to do something. We hear that you are going to Mount Wilson. When you get there, why don't you use the big telescope to make measures of stars in globular clusters?" Few besides Bailey found these stellar swarms appealing, and Bailey himself lacked access to a telescope large enough to probe them deeply, either from Cambridge or from Arequipa.

As often as Shapley could gain his own observing time on Mount Wilson's 60-inch reflecting telescope, he did as Bailey requested. "Within a month or two after I got to Mount Wilson," Shapley recorded in a memoir, "Shapley and the globular clusters became synonyms." In the clusters he found new examples of Miss Leavitt's stars. Soon he developed a theory about their nature: The Cepheids were not closely orbiting pairs, as most astronomers believed, but enormous, isolated individuals. He could claim as much because their sudden surges in brightness seemed to indicate outbursts of some kind, not a pattern of eclipses by a partner star. The Cepheids owed their variation, Shapley hypothesized, to dramatic pulsations in temperature and diameter. He described them as "throbbing or vibrating masses of gas."

A number of new Cepheids found by Shapley displayed the long periods indicative of great brightness according to Miss Leavitt's law. These gave Shapley a way to ascertain their positions and the distribution of the clusters in space. He adapted the techniques Ejnar Hertzsprung had used for deducing distances from period and magnitude, and began measuring the way out to some of the hundred clusters he could see. He noticed a group of them—a "cluster of clusters"—huddled together in a section of the Milky Way, near the constellation Sagittarius, the Archer. He wondered what was so special about that particular area.

• • •

THE OUTBREAK OF THE GREAT WAR in Europe hobbled astronomy there and placed a new imperative on the United States, as Pickering saw things, to sustain all fields of scientific research. He found himself in what seemed an ideal position to aid fellow scientists, as the recently appointed executive chair of the new Committee of 100 on Research, established in 1914 by the American Association for the Advancement of Science. His efforts got off to a bad start, however, when his early official appeals on behalf of four physics and astronomy projects failed to win funding from the Carnegie Institution.

As for the needs of his own institution, Pickering never tired of reminding President Lowell that Harvard contributed nothing to the observatory's support. Fear of fire still fueled the director's oft-repeated plea for more brick structures to replace the wooden ones, and he sighed with relief every time a manuscript series of observations found permanence in a published volume of the *Annals.* For several years up to and including 1914, expenses had exceeded income, forcing Pickering to curtail activities. In December 1914, however, the observatory received a large new sum under regrettable circumstances.

"During the last year," Pickering wrote in his annual report, "the Observatory has lost, by the death of Anna Palmer Draper, its most generous benefactor." She had died at home in New York, of pneumonia, on December 8. "It rarely happens that a woman maintains for many years a great scientific enterprise, and by monthly subscriptions shows her unfailing interest in it. Mrs. Draper for nearly thirty years supported the Henry Draper Memorial, and by her will placed it upon a permanent foundation."

The terms of Mrs. Draper's will promised the observatory the sum of $150,000, in addition to the quarter million she had already given, "for the purpose of caring for, preserving, studying and using the photographic plates of the Henry Draper Memorial." Anticipating a lengthy probate, Mrs. Draper had instructed her executors to pay the observatory $4,000 the year of her death and $5,000 every year thereafter until her estate was settled, so the work of the memorial could continue uninterrupted. An article in the *New York Times* on December 20 about her several bequests mistakenly identified her husband as a former professor of astronomy at Harvard. She had willed her country house in Dobbs Ferry to a nephew on the Palmer side, and although she left legacies to niece and nephew Carlotta and Draper Maury, she made no allowance for their older sister, Antonia.

Miss Cannon wrote a lengthy obituary notice for *Science,* in which she compared Mrs. Draper to another woman of her acquaintance, Lady Margaret Huggins. "It is interesting to note that the wives of two of the men connected with the beginnings of this science played such important parts in the careers of their husbands. . . . For Mrs. Draper not only was her husband's associate in his investigations during the fifteen years of their lives together, but after his early death in 1882, she was able to provide for carrying on his work in a most efficient manner." Miss Cannon did not mention her own part in the continuation of that work, except to say, "In 1911, observations were commenced for a New Draper Catalogue, which will contain the spectra of at least 200,000 stars situated over the entire sky. In this work Mrs. Draper was greatly interested until the very last, and wrote encouragingly about its progress."

When Lady Margaret Huggins died a few months later, in March 1915, Miss Cannon wrote her obituary, too. Eddington of the Royal Astronomical Society accepted the notice for publication in the *Observatory.* "It gives me quite a new appreciation of her personality," he said in his thanks to Miss Cannon on July 3. In the same letter, Eddington, a Quaker and therefore a pacifist, lamented the violent turn of current events: "It is very sad, after the jolly days in Bonn, that this division should have come between us and our German colleagues." The zeppelin that had so delighted the visiting astronomers in 1913 had become a force of destruction, dropping bombs on Great Britain. "If only there was mutual respect between the combatants, it would be a less depressing outlook," Eddington wrote, "but I am afraid the contempt and hatred of Germans has increased over here very much in the last 3 months, though personally I have not yet got to the length of imagining, say, Max Wolf, as a 'pirate and baby-killer.' The knowledge that we have the sympathy of nearly all American astronomers is much appreciated, because you have so much more opportunity than we have of learning what is to be said in favour of the other side."

Pickering, too, rued the war's undoing of the friendly ties binding observatories around the world. Already the usual course of international communication regarding comets and asteroids had been disrupted. Copenhagen took the place of Kiel, Germany, as the designated European clearinghouse for such information, but most astronomers on the continent were cut off. Even the cables to Copenhagen became problematic, when military censors on both sides of the Atlantic refused to permit the use of ciphers—over the protests of astronomers, who had always coded their messages (by substituting words for digits) to avoid errors in transmitting long strings of numbers.

PLEASED WITH MARGARET HARWOOD'S three productive summers on Nantucket, the Maria Mitchell Association granted her a quadrennial year, beginning June 15, 1915, to spend as she liked while still collecting her stipend of $1,000. She chose to go West, to assist at the Lick Observatory on Mount Hamilton while taking her master's degree in astronomy at the University of California, Berkeley.

"Dear Miss Cannon," she wrote June 23 on Lick stationery, "This letter is really to both you and Professor Pickering." She had too much to say to try to say it all twice.

"The trip out was perfect in every way." Miss Harwood had stayed with Edwin and Mary Frost at the Yerkes Observatory in Williams Bay, Wisconsin, and been entertained by Percival Lowell's staff in Flagstaff. In the Pasadena office and workshop of the Mount Wilson Observatory, she had met Harlow Shapley.

"I was having a good time talking variables with Mr. Shapley that first afternoon," Miss Harwood reported, "when his telephone rang and he answered by asking whoever wanted him to call up again in half an hour. I objected but he said it was nothing important and we went on. In about three quarters of an hour more the telephone rang again and I started to go. He called me back saying that Mrs. Shapley was just calling to find out if I was coming to dinner with him. She had called before, but he was too busy talking to ask such an everyday question! So I went and had a delightful time. Mrs. Shapley is very attractive and young, and, if possible, more of a 'shark' than her husband." Shapley had married his college sweetheart, Martha Betz, in Kansas City in April 1914, then boarded the train with her for a honeymoon trip to their new home in Pasadena.

"She was working for her philology degree at Bryn Mawr when she married," Miss Harwood continued. "She plays the piano finely, tends the three months old baby girl (who is also a wonder and most attractive) and is a wonderful cook. Mr. Shapley has taught her astronomy so that she measures plates, works out the light curve of variables and writes out the discussion by herself. She is very shy and retiring and I found out almost nothing about her until I questioned Mr. Shapley on the way home [to the boardinghouse where several Mount Wilson computers lived]. She did play to me, however."

The following day saw Miss Harwood ascend the winding road to the Mount Wilson summit, where she overnighted. "I stayed up until 1:00 AM watching Mr. Shapley photograph certain clusters. And I took a plate on Messier 3 with the 60 inch! I have not seen the plate yet and so better not be bragging. At 12 o'clock Mr. Hoge, the night assistant, makes cocoa in the kitchen of the 60-inch dome and we had a regular feast of strawberries, cocoa, toast and pilot bread. As each observer works all night, the midnight meal is very necessary and must go to the right spot."

Miss Harwood's big news for her mentors at Harvard concerned a forwarded letter that she had found waiting for her in Berkeley. It came from Ellen Fitz Pendleton, the president of Wellesley College, offering her a position as an instructor at "a salary of not less than $1,200" beginning in the 1916–1917 academic year, with the strong possibility of promotion and a pay increase in 1917–1918. The timing would allow her to complete her master's degree, but she needed to make the Wellesley decision now.

Miss Cannon might well have thrilled at the prospect of her protégée on the faculty of her alma mater, but she balked at the thought that teaching would take Miss Harwood away from her ongoing research regarding the light curve of the asteroid Eros. Lydia Hinchman at the Maria Mitchell Association, who had been a teacher herself, thought abandoning research a terrible idea.

"I do not want to see her exchange the life of an astronomer for that of a teacher," Mrs. Hinchman insisted to Miss Cannon on September 7, 1915. "A teacher grows weary and old before her time, but if I may judge from you my dear Miss Cannon, an astronomer is always young."

Acting quickly, Mrs. Hinchman proposed that the association's board designate Miss Harwood as the permanent fellow as well as director of the Maria Mitchell Observatory, with a salary to match the Wellesley offer. Her plan drew strong resistance from board member Anne Sewell Young of Mount Holyoke College. "While I value highly the very excellent work which Miss Harwood has done at Harvard as well as at Nantucket," Dr. Young argued, "and appreciate the tact and good judgment which have so endeared her to the residents of Nantucket, I cannot approve of appointing a permanent Fellow as Director of the Nantucket Observatory. I strongly believe that the Cause of Astronomy as well as that of woman's education will gain far more by continuing to offer one or more fellowships, giving opportunities for study and research to various women of ability or promise.... Those of us who are teaching know how few, even now, are the opportunities offered to women, and have been proud of this fellowship in astronomy. Miss Mitchell's interest in 'her girls' was so great that it seems to me she would herself choose this for her memorial. I am very certain that my opinion is that of Prof. [Caroline] Furness of Vassar who did so much toward raising the fund, also of Prof. [Harriet] Bigelow of Smith, and Prof. [Sarah] Whiting of Wellesley College."

Mrs. Hinchman bristled. As a blood relative of Miss Mitchell, she did not appreciate outsiders' interpretations of the deceased astronomer's wishes. Moreover Mrs. Hinchman and her husband, Charles, had contributed by far the largest sum and greatest effort toward the establishment of the fellowship. She steeled herself to sway the board toward her opinion. "Meeting set for October 6 at the College Club," she informed Miss Cannon, head of the fellowship committee. "I think they should know about Miss Harwood's work, and experiences in the west. . . . I shall too make it as plain as I courteously can, that our observatory was meant for research—It was in no sense intended as a training school for teachers and the height of efficiency seems to me to have been reached, when its advantages provide the opportunity for a Fellow to complete undertaken work." The other members voted along with her, and thereby made Miss Harwood, who happily accepted the directorship, the only woman in the world in charge of an independent observatory. She was thirty years old, the same age as Pickering when he took over at Harvard.

No sooner had Mrs. Hinchman won her way on Miss Harwood's account than she saw the virtue of creating a second astronomical fellowship at Nantucket. She put together a committee that spent a year soliciting funds from Nantucketers, Harvard friends, and former students of Maria Mitchell. On November 16, 1916, in the Harvard Observatory, distinguished Vassar alumna Florence Cushing handed Pickering a check for $12,000 as an added surprise at the big surprise party celebrating his fortieth year as director. "It is our wish," Miss Cushing told him, "that you will accept it with full power as to its use, and that in the future it will be managed with the same broad-minded fairness to women which has characterized your administration."

The committee had thought of calling the second stipend the Harvard Fellowship, but President Lowell pointed out that the university's name could not be attached to a fund controlled by an individual. At that, the Maria Mitchell Association moved to rename its new annual award the Edward C. Pickering Astronomical Fellowship for Women.

"PRESIDENT WILSON HAS SEVERED diplomatic relations with Germany," Miss Cannon wrote in her diary on Sunday, February 4, 1917. "The terrible submarine warfare is on again." Pickering had discussed the submarine threat with the chairman of the U.S. Navy's Consulting Board, Thomas Edison, at the start of the hostilities, suggesting strategies and offering all the resources of the Committee of 100 on Research. After the United States declared war on Germany in April 1917, Pickering bent his inventive mind more urgently to military needs. Together with Willard Gerrish, the observatory's resident mechanical genius, he devised a means for heavy artillery operators to orient their equipment. The new device, like Pickering's early photometers, relied on sighting the North Star. The War Department welcomed his model of the "Harvard Polaris Attachment," and informed him of its plans to produce the instrument.

At Mount Wilson, Harlow Shapley announced his plans to enlist in the coast artillery, but Director George Ellery Hale advised him not to, on the grounds that he would likely be needed to help with crucial optical projects for the National Research Council. Shapley agreed to stay in Pasadena for the time being, keeping up his observations of clusters and Cepheids.

"Very much of my work on clusters," Shapley wrote to Bailey on January 30, 1917, "has been the direct result of my conversation with you in Cambridge three years ago when you suggested the advantages of the Mount Wilson instruments and weather and when you expressed the hope that I would join in the study." Since then Shapley had determined the distances to all the clusters that contained Cepheid-type variable stars, thanks to the period-luminosity relation. In so doing, he had assumed that Miss Leavitt's law was not limited to the Magellanic Clouds, but could govern conditions anywhere.

To place the clusters *not* containing Cepheids, Shapley combined a variety of means and assumptions to leapfrog his way across space. Often he relied on other cluster variables that were faster-paced than Cepheids, but which seemed also to obey Miss Leavitt's law. In August he wanted to pursue this line of thought with her, but she was away on vacation in Nantucket, visiting Miss Harwood.

In clusters too distant to reveal any variables, Shapley averaged the magnitudes of the thirty brightest stars he could espy. He then compared these averages with the average luminosity of the Cepheid-containing clusters, and deduced the greater distances accordingly. For clusters too far removed to be resolved into any stars at all, Shapley measured the one thing he could see—their apparent overall diameters, which he compared with the diameters of clusters whose distances he had already determined.

Shapley gauged the average cluster diameter to be about one hundred fifty light-years, or nine hundred trillion miles across. The range of cluster distances from the Sun reached the still more staggering numbers of fifteen thousand to two hundred thousand light-years. No other astronomer had stretched the bounds of the known universe to those extreme dimensions.

A chance encounter with ants on Mount Wilson briefly turned Shapley's attention from the very big and distant to the small and close at hand. While watching patrols of trail-runner ants traverse the back of a concrete building, he noticed that they slowed from a run to a walk as they passed through the shade of the manzanita bushes. At first he presumed the ants were enjoying a cool respite, just as he was doing at that moment. "I began to wonder about this, however," he recalled in his autobiography, "and soon I got a thermometer and a barometer and a hydrometer and all those 'ometers,' and a stop watch. I set up a sort of observing station while resting and getting ready for another night's tussle with the globular clusters." Shapley found that the ants set their pace by the ambient temperature. The higher the mercury, the faster they ran, even when carrying loads. No other factors, such as air pressure or humidity, affected their rate of travel. "I found it great fun to watch them." Recording his ant observations as carefully as he treated any other scientific phenomenon, Shapley arrived at a temperature-speed ratio. He clocked ants in 35-degree cold (*Liometopum apiculatum* on the edge of a snow bank) and 103-degree heat (*Tapinoma sessile* in the den of Shapley's Pasadena home, where he stripped off his clothes and turned up the thermostat to test the limit of the ants' tolerance). He claimed he could tell the air temperature to within one degree by watching half a dozen ants pass through his "speed trap," and he published his data on ant thermokinetics in the *Proceedings of the National Academy of Sciences.*

In time Shapley came to see the globular clusters as the scaffolding of the cosmos. Scores of clusters attended the Milky Way. Some hovered above the broad flat plane of stars, while others dipped below it. Together, they described an enormous halo enveloping the entire galaxy. Shapley could tell by the distribution of the clusters that his own vantage point—atop Mount Wilson, riding planet Earth around the Sun—stood nowhere near the center of the Milky Way. Were he situated at the galactic center, he reasoned, he would see the clusters evenly spaced around him. As it was, when he looked in one direction he saw a sparse string of clusters, and in the opposite direction the "cluster of clusters" in Sagittarius. He concluded the center must lie in that direction. The Sun, though the center of the solar system, was not the hub of the universe. "Some of my series of papers on globular clusters published in 1917 and 1918 were rather revolutionary because the findings opened up a part of the universe that had not been known before," he wrote of his bold speculation. In Shapley's new cosmic picture, "the solar system is off center and consequently man is too, which is a rather nice idea because it means that man is not such a big chicken. He is incidental—my favorite term is 'peripheral.'"

There was no telling how far into space astronomers might penetrate by the light of Miss Leavitt's stars. Having limned the extent of the Milky Way on a foundation of Cepheids, Shapley recognized the need to refine Miss Leavitt's magnitude measurements, to make sure they were strong enough to support his conclusions. In a letter to Pickering on July 20, 1918, Shapley stated, "I believe the most important photometric work that can be done on Cepheid variables at the present time is a study of the Harvard plates of the Magellanic clouds. Probably Miss Leavitt's many other problems have interrupted and delayed her work on the variables of the clouds for the interval of six or seven years since her preliminary work was published." No doubt her illness, which had been diagnosed as cancer, figured chief among Miss Leavitt's problems, though her many other scientific assignments had effectively barred her from further pursuit of her Cepheid discoveries.

Shapley closed his letter with a prediction: "The theory of stellar variation, the laws of stellar luminosities, the arrangement of objects throughout the whole galactic system, the structure of the clouds—all these problems will benefit directly or indirectly from a further knowledge of the Cepheid variables."

THE MEMBERS OF THE AAVSO, those devoted observers of the long-period variables, met in November 1918 at the Harvard College Observatory. They had been accustomed to getting together in Connecticut or New Jersey at the homes of the association's officers, but now that Leon Campbell had returned from Peru and resumed close communication with the volunteers, the observatory served as the new unofficial headquarters. To further cement the Harvard connection, the organization inducted Solon Bailey, Annie Cannon, Henrietta Leavitt, and Edward Pickering as honorary members, with a special tribute for the director: "He has assisted us in everything that we have undertaken, and has carefully watched our progress along every step of the way." Founder William Tyler Olcott compared Pickering's manner to that of a benevolent older brother.

In 1918 part one of the long-awaited revision of the Henry Draper Catalogue by Annie J. Cannon and Edward C. Pickering appeared in print. Pickering personally covered the cost to publish it as volume 91 of the *Annals,* and described the process of its preparation in his preface. Beyond Miss Cannon's four years of "unfailing enthusiasm" and "persistent work" in reclassifying the spectra of 222,000 stars, she had also invested two years in writing remarks and otherwise readying the material for printing. At least five assistants at a time, though not always the same five, had aided her throughout. Pickering named "Misses Grace R. Brooks, Alta M. Carpenter, Florence Cushman, Edith F. Gill, Mabel A. Gill, Marian A. Hawes, Hannah S. Locke, Joan C. Mackie, Louisa D. Wells, and Marion A. White" as the ones who had ascertained the positions and magnitudes of every star included, as well as helped proofread the hundreds of pages of tables and text. He stressed the efficiency of the women's effort and cooperation: "A loss of one minute in the reduction of each estimate would delay the publication of the entire work by the equivalent of the time of one assistant for two years."

By this point Pickering, who liked to keep count of everything, figured the observatory had satisfied thirty-seven thousand outside requests for spectral classifications. Projecting the frequency with which astronomers would consult the printed authority far into the future, he took pains to choose paper that "should not be affected by time." Although experts had assured him a rag content of sixty percent would be more than adequate, he opted for eighty instead, despite the added expense. "It is hoped that these volumes"—eight more were to follow the current issue—"will form a lasting tribute to the memory of both Dr. and Mrs. Draper."

A tipped-in frontispiece illustrated the principal types of spectra, B through M, with their distinguishing features, but both Pickering's preface and Miss Cannon's coda apologized that these reproductions portrayed only a fraction of the Fraunhofer lines visible in the original glass negatives.

For the second volume of the catalogue (*Annals* volume 92), which came out later the same year, the authors chose a studio portrait of Henry Draper as the frontispiece. It depicted the doctor in two-thirds facial view, with an expression serious but not stern, and a few stray locks of hair jutting out by his ear. The same photograph had served as the model for the Henry Draper gold medal awarded by the National Academy of Sciences.

On Christmas Day 1918, Pickering wrote a brief preface for installment number three of the catalogue series. "The Henry Draper Memorial is due to the unfailing devotion of Mrs. Draper to the memory of her husband," he reflected. "It seems, therefore, very appropriate that her portrait should appear as a Frontispiece to this, the third volume of the greatest work yet undertaken as part of the Henry Draper Memorial." Mrs. Draper, seen here in profile, looked ready to welcome guests at one of her Academy dinners. She wore a gown with exquisite lacework and her red hair pinned up in tight ringlets.

• • •

WELL BEFORE THE WORLD WAR ENDED, Pickering began agitating for the resumption of international relations among scientists. In August 1918 he had told George Ellery Hale, "No ordinary punishment is adequate for those responsible for barbarities contrary to the laws of nations and humanity, yet we ought not to ignore the work of those who, laboring quietly in their observatories, have done their best to extend our knowledge in these terrible times." After the Armistice of November 1918, Pickering declared himself eager to write to longtime friends in Germany as soon as mail communication could be restored. "I am anxious to know how far European observatories have suffered," he said in a letter to Elis Strömgren of the Østervold Observatory in Copenhagen on January 7, 1919, "and what is likely to be their condition when the peace treaties are actually signed." He resented the sentiments of certain colleagues in the United States and the United Kingdom who talked of excluding scientists of enemy or neutral nations from postwar professional societies. "I believe that many astronomers agree with me," he told Strömgren, "that we should make every effort for the advancement of our science, regardless of personal or national considerations."

Later that month, however, Pickering's own efforts came to an end. His strength suddenly failed him while he was working at the observatory, and he needed to be helped over the few steps home to his residence. The cause of his death, on February 3, was given as pneumonia.

Edward Pickering had been director of the Harvard College Observatory for forty-two years, serving longer than the combined tenures of all his predecessors. Anguish at his loss was widespread.

"I warmly admired his great ability, his originality of view, his power of organization, and his unwearied initiative," George Ellery Hale wrote to Solon Bailey on February 4. "I also appreciated how much he did, in so many ways, to stimulate research and to help astronomers everywhere. The great development of the observatory under his direction, and its immense contribution to the progress of astronomy, mark an epoch, universally recognized, in the advancement of science." Hale, who had volunteered as a Harvard assistant while a student at MIT, said he still remembered Pickering's showing him the original photographs of stellar spectra made by Henry Draper. "What I shall remember with greatest pleasure, however, is his kindly interest in me as an unknown amateur when I first came to the observatory. Many others have enjoyed this experience, for the circle of amateurs he touched and helped was a wide one."

Bailey, like Hale, had first arrived at the observatory as an amateur volunteer. One of his first duties now, as acting director in the great man's absence, was to compose Pickering's obituary for Hale's *Astrophysical Journal.* "For men and women he had an equal charm," Bailey commented after elaborating the milestones of the illustrious life story. "His grace of manner and conversation was the constant wonder of all who knew him intimately. Over all, old and young, wise and witty or ignorant and stupid, who seemed to have any claim upon him, he threw the glamor of his personality."

Bailey spoke also of the library of glass plates, gathered pole to pole and year to year, as the repository of Pickering's guiding spirit: "It still exists, its possibilities by no means exhausted, its value in many ways increasing as the years go by.... Within this great collection of stellar photographs... there still remains for Professor Pickering the possibility of an immortality of scientific labor, more unique and worthy than ordinary fame."

Miss Cannon, the consummate master of obituary form, sketched the director's much admired qualities in *Popular Astronomy:* "He will be missed for his warm-heartedness, always eager to help the young astronomer, whether by securing grants of funds or in the selection of his life work; for his cordiality, the ideal host in welcoming visitors to the Observatory; for his sympathetic, inspiring personality, which, by its very optimism and faith in humanity, made us believe in ourselves and our capabilities."

Miss Cannon concluded, "His joy in taking part in what he called the greatest problem ever presented to the mind of man, the study of the starry universe, never left him, and, even in his last illness, he spoke of having new ideas about work. . . . He measured the light of the stars and first placed them in an orderly evolutionary sequence. He left, as his legacy to the world, the history of the sky for the last thirty-five years imprinted on the Harvard collection of photographs."

PART THREE

In the Depths Above

I saw in the stars a chance to observe phenomena beyond terrestrial scope. Nothing seemed impossible in those early days; we were going to understand everything tomorrow.

—Cecilia Payne-Gaposchkin (1900–1979)
Phillips Astronomer, Harvard College Observatory

There are two ways of spreading light: to be the candle or the mirror that reflects it.

—Edith Wharton (1862–1937)
Author of *The Age of Innocence* and other acclaimed novels

CHAPTER ELEVEN

Shapley's "Kilo-Girl" Hours

MARY H. VANN, an alumna of Cornell University, devoted her 1917–1918 term as the first Edward C. Pickering Fellow to an analysis of the new stars, or novae, that had shown up on the Harvard glass plates since 1887. Most of the eleven such stars had never been observed visually by anyone, or photographed by any other institution. Now, thanks to the abundance of plates and Miss Leavitt's completion of the North Polar Sequence, Miss Vann had the necessary tools to assess the novae's changing magnitudes over time and create a light curve for each one. On June 8, 1918, shortly before she left the observatory to take up war work, a *new* nova erupted in the constellation Aquila, outshining all but the very brightest stars for several weeks. At magnitude –0.5, Nova Aquilae 1918 proved the brightest such sight since the invention of the telescope, but its photographic study fell to the second Pickering Fellow, Dorothy W. Block, a 1915 graduate of Hunter College in New York City.

Unlike the astronomical fellowship of the Nantucket Maria Mitchell Association, now permanently assigned to Margaret Harwood, the Pickering Fellowship entailed no Nantucket residency rule. The recipient was welcome to visit Miss Harwood on the island during the summer months, if she so chose, but the real reward consisted of research funding at Harvard through a typical fall-to-spring academic year. Miss Block devoted her allotted time in 1918–1919 to measuring the changing light of variable stars, several asteroids, and of course the Great New Star in Aquila. In the spring, she also learned to photograph the stars, so that she often stood in for the regular assistant during the first part of the night. This experience helped her secure a job offer from the Yerkes Observatory in Williams Bay, Wisconsin, where she was about to become the first woman ever allowed to take photographs through its 40-inch telescope, the world's largest refractor.

As Miss Block prepared to leave Cambridge, Henry Draper's niece Antonia Maury petitioned to become the next Pickering Fellow. At fifty-three, the experienced Miss Maury was roughly twice the age of either Miss Vann or Miss Block. However, she met the crucial requirement of holding a college degree, Vassar class of 1887, and had studied astronomy under the beloved Maria Mitchell.

"With regard to Miss Maury," Lydia Hinchman of the Maria Mitchell Association asked Annie Cannon in a letter of April 8, 1919, "I have heard she is peculiar, is that true? I also have heard she has a brilliant mind. I scarcely feel competent to give an opinion without seeing her, but it is for one year and if you feel so disposed, you might try it." Miss Cannon chaired the selection committee, but Mrs. Hinchman was entitled—and accustomed—to offer advice: "If her peculiarities are not to the front I would try it."

Miss Maury had resumed her on-again, off-again association with the Harvard Observatory in August 1918, by attending the American Astronomical Society meeting held there. It was her first meeting as an elected member of the organization. Pickering, who was then the society's president, had invited Miss Maury to stay on in Cambridge as a voluntary associate researcher. Thus, with no prospect of salary, she rekindled her first love—the very close double stars known as spectroscopic binaries. Months later, following the director's death, Professor Bailey encouraged Miss Maury to take a paid position as assistant to John Stanley Plaskett at a brand-new observatory in Victoria, British Columbia, but, at this stage of her life, she could not readily relocate so far from anywhere she had ever called home.

Miss Cannon and Miss Maury, close contemporaries, had overlapped in their careers at Harvard long enough to judge each other's character and quirks. Miss Cannon deemed her colleague entirely deserving of the fellowship opportunity, and Miss Maury gratefully accepted the $500 stipend.

THE FIRST TWO PARTS of the revised and expanded Henry Draper Catalogue, published as *Annals* volumes 91 and 92, made Harlow Shapley impatient for the third. "Can you predict when 93 will be distributed?" he inquired of Miss Cannon from his post at Mount Wilson on May 8, 1919. "It happens to be the most important of all to me. I am using your results to check my work on cluster structure, and the stars of the southern Milky Way . . . play an important role." By "cluster structure" Shapley referred not to the distant globular clusters ringing the periphery of the Milky Way, but to what he termed the local cluster, meaning stars in the vicinity of the Sun—stars close enough to be described in the catalogue according to their position, magnitude, and spectrum. Parts one and two of Miss Cannon's magnum opus had covered several longitudinal swaths of the 360-degree sky panorama, from zero degrees, or the "zero hour" in astronomical parlance, to ninety degrees, the sixth hour. Shapley now needed the seventh and succeeding hours, yet to be revealed in the next installments, to continue his study of the organization of the solar neighborhood.

Volume 93, Miss Cannon assured him, had already been set in type, but the bookbinders had gone on strike, and she could not say how long publication might be delayed by the settling of their grievances. Meanwhile she satisfied Shapley by mailing him the unbound signatures. The observatory, even in Pickering's absence, was continuing to operate by his guiding principle: first you gather all the information, then you give it away to those who crave it.

"My very best thanks for your kindness in sending me proof-sheets of the third volume of the Henry Draper Catalogue," Shapley wrote back. "I have been through it all and have obtained just the information I thought would be there relative to the form and extent of the local cluster." Shapley wanted more than one way to measure distances across the galaxy. The Cepheids had demonstrated their special powers as distance indicators, but

Miss Leavitt's stars were few in number. Shapley thought the more numerous, luminous stars belonging to spectral class B could also provide distance clues. B-type stars were generously scattered through the Milky Way, and their positions and magnitudes already well established through thousands of reliable measures, all collated in the Harvard catalogues. In general, the B stars shone some two hundred times brighter than the Sun. From Mount Wilson, Shapley could make out the spectra of B stars among distant clusters via the 60- and 100-inch telescopes. By the relative faintness of the remoter B's, he was able to estimate their distance, and thus employ them as alternate mileposts. Shapley thought the giant red stars could also be engaged as measuring aids, since they, too, populated both the globular clusters and the Milky Way.

Many other researchers, pursuing other studies, echoed Shapley's clamoring for the remainder of the Henry Draper Catalogue. The later volumes, however, faced a problem more serious than labor disputes, namely lack of funds. "The prompt publication of all this material is necessary in order to complete the great life work of the late Director," Solon Bailey stressed in his first report as acting director. He projected the cost at $15,000 beyond the income of the observatory. While seeking funding, he honored and tallied the urgent requests from individual astronomers for particular spectra. Hundreds of such queries arrived every month.

Edward Pickering had told President Lowell in 1910 that he considered Professor Bailey the only member of his staff capable of taking over the observatory as acting or permanent director. After Pickering's death in 1919, Bailey's assumption of responsibility had proceeded seamlessly, but the Harvard administration made no move to confirm him as the fifth director. George R. Agassiz, a member of the Visiting Committee and patron of the observatory, counseled Lowell to go for "new blood and real distinction." Not even Bailey considered himself, at age sixty-five, the right person to lead the observatory into the future. He pictured a younger man taking charge, someone like Harlow Shapley of Mount Wilson—or better yet Shapley's mentor, Henry Norris Russell of Princeton, only forty-two years old and widely regarded as a brilliant thinker. On learning of his eligibility, the cautious Russell raised an eyebrow. He half-suspected that Abbott Lowell would have appointed his own "distinguished brother," the Mars expert, as director, if only Percival Lowell were still alive. The founder of the Lowell Observatory, however, had passed away at Flagstaff in 1916. Pickering's brother, William, on the other hand, though not in the running for the directorship, remained at large in Mandeville. Should Russell accept Harvard's overtures, he would inherit William along with the other staff members, and that thought gave him pause. William seemed obsessed with Martian canals, claimed to have detected water on the Moon, and was known to be calculating the whereabouts of a planet beyond Neptune.

Neither Miss Cannon nor Miss Leavitt, being female and fifty-plus, was eligible for the directorship. Nor did either wish the position for herself. Miss Leavitt, never physically strong, had been forced to leave the big house on Garden Street when it was sold after her uncle Erasmus Leavitt died in 1916. She had moved into a rooming house, but when her widowed mother came back East, the two of them took an apartment together on Linnaean Street, close to the observatory. Miss Cannon, still living happily with her older half sister, Ella Cannon Marshall, continued to accrue honors at home and abroad. The University of Delaware awarded her a doctor of science degree in 1918, claiming her as a distinguished native daughter of the "Diamond State." In 1919 her English friend Herbert Hall Turner, the Savilian Professor of Astronomy at Oxford, sought to raise her standing in the Royal Astronomical Society. "The other day," Turner wrote her on May 13, "I proposed your name as an Associate of the R.A.S.—on the same footing as men. I hoped you would regard it as a new recognition: & that by thus transferring our one 'Honorary Member' we should remove the last remaining trace of the previous disability of women. But the Council disagreed with me & thought you would prefer your present 'lonely isolation' as a greater honor."

Thanks to a 1915 change in the society's charter, women could now be elected fellows (if British subjects) and associates (if foreigners). Miss Cannon felt content to retain her "honorary" status, but she took Turner up on another of his suggestions in the spring of 1919, regarding the Maria

Mitchell Association. "It might be possible, let us say as a friendly act at the present moment of great events and important new departures," Turner thought, "to assign one of the Fellowships to an Englishwoman. I need scarcely point to the advantages which such action would have both in encouraging woman's work generally, in cementing friendly relations between the two nations, and in creating a new form of recognition." Miss Cannon's committee had already decided on Miss Maury for the coming year, but the members heard in Turner's words an echo of "Professor Pickering's international spirit," and promised to search overseas for the next Pickering Fellow.

HARLOW SHAPLEY'S "BIG GALAXY," as he described it in 1918, filled the known universe. It was so immense that it subsumed everything else: globular clusters rimmed it, nebulous forms fit inside it, and the Magellanic Clouds hung from it as appendages. But numerous astronomers refused to be bound by it. Unlike Shapley, they viewed the Milky Way as one galaxy among many—a single "island universe" in a vast archipelago.

Shapley, too, had ascribed to the island-universe theory up until 1917. But once he exploded the size of the Milky Way to titanic proportions through his distance assessments of the globular clusters, he changed his mind. The vastitude of the Milky Way ruled out the existence of other, comparable galaxies. Nothing substantive surrounded it, Shapley thought, other than dross and empty space.

Deciding the truth of the island-universe matter depended on determining the placement of the spiral nebulae. These pinwheels of celestial light had been sighted by the thousands, beginning in the mid-nineteenth century, when William Parsons of Ireland and his friends first viewed their distinctive forms through the huge reflecting telescope known as the Leviathan of Parsonstown. The spirals, as they were called for short, looked as though they might be whirlpools of incandescent gas, or whirlwinds of interstellar dust, or whorls of stars. It was difficult to say without knowing their distances. Some astronomers saw each spiral's bright center and trailing arms as a new solar system of a sun and planets in the making. Those who viewed the spirals as full-blown external galaxies, however, divined in their coiled forms a likely blueprint for the Milky Way.

George Ellery Hale thought the disagreement over the spirals an apt subject for a public debate. When he proposed the topic to the National Academy of Sciences late in 1919, he also named general relativity, which had been much in the news, as an alternate theme possibility. The idea of relativity put forth by Albert Einstein in 1915 was changing the nature of space from a passive container of the stars to a fabric warped by the stars' presence. Einstein's German roots and the course of the Great War at first slowed the theory's acceptance, but the English pacifist Arthur Stanley Eddington tested its validity during the May 29, 1919, total solar eclipse, which he observed from the African island of Príncipe. It was an eclipse expedition even Pickering would have approved. The stunning results, announced in November 1919, demonstrated that light waves indeed felt the effect of gravity—and by the amount Einstein had predicted. The erudite Eddington expressed the findings in poetry as well as prose, borrowing the rhythm of the *Rubaiyat of Omar Khayyam:* "Oh leave the Wise our measures to collate / One thing at least is certain, light has weight / One thing is certain and the rest debate / Light rays, when near the Sun, do not go straight."

Given a choice between relativity and the galaxy, the Academy secretary, solar astronomer Charles Greeley Abbot, expressed a strong preference: "As to relativity I must confess that I would rather have a subject in which there would be a half-dozen members of the Academy competent enough to understand at least a few words of what the speakers were saying if we had a symposium upon it. I pray to God that the progress of science will send relativity to some region of space beyond the fourth dimension, from whence it may never return to plague us." Having thus settled the discussion topic in favor of spirals, Abbot invited Shapley to present his mono-galaxy idea, and Heber D. Curtis of the Lick Observatory to argue the case for multiple galaxies.

The event took place in Washington, D.C., on the evening of April 26, 1920. Shapley, who was known to behave at times in brash and overconfident ways, withered even before he took the podium. Not only did he fear being upstaged by a smooth public speaker of Curtis's stature, but he had learned well ahead of showtime that Agassiz from the Harvard Observatory's Visiting Committee would be in the audience to judge his fitness for the director slot. Unfortunately, Shapley pitched his prepared talk at a level appropriate for intelligent laymen, and it impressed no one. Speaking first, he took several minutes to explain the meaning of a light-year as the distance that light traveled in a year. "Now that we have a satisfactory unit of sidereal distance," he said, reading from his script, "let us go rambling about the universe." He led a pictorial tour through star clusters near and far, including those seen in Orion and Hercules, but promised, "I shall not impose upon you the dreary technicalities of the methods of determining the distance of globular clusters." He sidestepped the spirals, except to stress how little was actually known about them. "I prefer to believe that they are not composed of stars at all, but are truly nebulous objects"—in a word, diffuse. Even if the spirals *are* stellar, he conceded in closing, they are *not* comparable in size with our stellar system, the Milky Way.

Curtis then came forward, intent on reducing Shapley's gargantuan galaxy to about one-tenth its grandeur—to the apparent size, that is, of a typical spiral. He laid out abundant arguments in favor of the spirals' being galaxies, including evidence for the Milky Way's own spiral shape. Examination of spirals' spectra, Curtis said, suggested that many more of them were made of stars than of free-flowing gas. In recent years, about a dozen spirals had lit up with new-star flare-ups, such as the 1895 nova that the late Williamina Fleming discovered in the Centaurus spiral. Curtis interpreted the presence of these novae as proof that spirals contained at least some stars, although opponents of the island-universe idea argued that novae arose when spirals collided with stars. Certainly spirals were on the move: Their spectra indicated tremendous speeds in the line of sight, as though most of them were rushing away from the Sun. Curtis saw these fantastic velocities as further testimony to the spirals' extragalactic location, since no stars within the Milky Way moved as rapidly. Curtis made all his points forcefully, and afterward boasted justly to his family that he had triumphed in the debate.

The face-to-face confrontation ended when the auditorium emptied that night, but the question of the spirals remained undecided. Shapley and Curtis continued their contest via correspondence over the ensuing months as they framed their presentations for publication in the *Bulletin of the National Research Council.* They traded drafts and weighed the worth of competing claims, but neither could win the other over to his side. While Shapley waited to hear whether Harvard would hire him, Curtis accepted the directorship of the Allegheny Observatory, and moved from California to Pennsylvania.

Those two states, California and Pennsylvania, had joined Massachusetts, Missouri, and thirty-one others by the summer of 1920 in ratifying the nineteenth amendment to the U.S. Constitution. One more state's sanction was still needed before women nationwide gained the right to vote. On August 18, during a special session in the Tennessee House of Representatives, the measure narrowly won support and became law. Miss Cannon went to the polls at the first opportunity, on September 7, to cast her ballot in the primary. She marked November 2, 1920, as "Election Tuesday" in her diary: "Gray day, and cold. Women out in force. I went with the Baileys at 10:30. Voting is very easy!" That evening she stopped on Boston Common to get the latest election bulletin, and absorbed the general enthusiasm for Senator Warren G. Harding of Ohio as the twenty-ninth president of the United States.

In England that autumn, the fourth Pickering Fellow, Miss A. Grace Cook of Stowmarket, stayed outdoors for hours every night to observe meteors, commonly known as falling or shooting stars. Seated in a deck chair, Miss Cook scanned the skies for sudden moving lights that signaled the entry of a bit of space rock or comet dust into Earth's atmosphere. When a meteor appeared she clicked her stopwatch to time its flight, and with her other hand held aloft a straight, thin wand, about five feet long, aligned with the intruder's path. In the few seconds of visibility, she memorized the meteor's changing magnitude compared with that of the stars it passed from apparition to disappearance, and then jotted down her accumulated data. In daytime she could plot the several pathways on a celestial globe to find the radiant, or point of origin, for a given meteor shower. And although the wet

English weather often thwarted her, she also observed other naked-eye phenomena such as aurorae and lunar halos, and hunted comets through the small telescope she purchased with her stipend. On February 9, 1921, after receiving the second half of her grant money from Professor Turner of Oxford, she wrote to Miss Cannon, "He seems to understand what a boon such a gift is to an isolated worker and one who can only set aside a small sum yearly to devote to science. It is like a lovely dream come true. I only hope I have made the very best use of it. I have done my best to do so." After months of reclusive work she added, "Most of my astronomical friends imagine I am in America at Harvard; they think the Fellowship was a residential one!"

MISS CANNON HAD KNOWN all along how things would turn out. The very first time she met Harlow Shapley, during his 1914 visit to Cambridge as a Princeton graduate student, she told him, "Young man, I know what you're going to do. You're going to be the director of the Harvard Observatory." Then she laughed. Shapley remembered her laugh years later, as something prophetic, or possibly psychic, when Harvard at last offered him the job.

"Dr. Shapley arrived!" Miss Cannon wrote in her diary on March 28, 1921. The next day she had "a long talk" with him, and decided, "I like him. So young, so clear, so brilliant." In fact, the thirty-five-year-old Shapley had not yet been named director; he was technically on probation, with the vague title of "observer." Given his sub-stellar performance in the scale-of-the-universe debate, and the hubris with which he put forth his daring theories, the untried new leader had been given just one year to prove himself worthy of Harvard's trust. If he should clash with the university or the observatory, then George Ellery Hale would gladly take him back at Mount Wilson.

Shapley himself considered the Cambridge relocation a permanent one. He spent several weeks that spring preparing the director's residence to receive his family, while Martha and the three children, Mildred, Willis, and Alan, visited relatives in Kansas City.

On his first day at the observatory, Shapley stopped by Miss Cannon's office and asked to see the spectrum of SW Andromedae, a faint variable that had aroused his interest. She called out to an assistant to retrieve a particular plate, identifying it by a five-digit number catalogued in her prodigious memory. To Shapley's amazement, "The girl went to the stacks and got the plate and SW Andromedae was on it!"

With Miss Cannon, Shapley initiated an investigation into the distribution of the stars of different spectral types, tallying the number in each class over a wide range of magnitudes. Pickering had attempted a similar statistical analysis decades before, with only one-twentieth the amount of data Shapley now found at his disposal in the Harvard plate collection. The Brick Building held the whole sky captive inside a hive of industry.

"Luckily Harvard College was swarming with cheap assistants," Shapley said of his early days as observer. "That was how we got things done." At Mount Wilson he had grown accustomed to making his own measurements of photographic plates. At Harvard he invented the term "girl-hour" for the time spent by young and not-so-young women at various plate-measuring and computing tasks. "Some jobs," he quipped, "even took several kilo-girl hours." Surely the ongoing preparation of the Henry Draper Catalogue figured among the more laborious. The fourth volume had been printed before Shapley's arrival, with the help of donations from observatory friends James and Margaret Jewett and members of the American Association of Variable Star Observers. Now veteran computer Florence Cushman was reading Miss Cannon's proofs for volumes five and six.

Shapley passed over the prim Ida Woods, who had served as Pickering's unofficial secretary, and instead selected the younger, more affable Arville "Billy" Walker to assist him with his correspondence. He straightaway engaged Miss Leavitt in a study of the different types of variables in the Magellanic Clouds. Together they demonstrated that the Clouds contained short-period, cluster-type variables, in addition to Cepheids. This was just the confirmation Shapley needed to shore up the great distances he had derived for the globular clusters—the distances on which his enlargement of the galaxy depended.

Further support for Shapley's Big Galaxy reached him in the spring of 1921 from his Mount Wilson friend and colleague Adriaan van Maanen. After comparing plates of the same spirals taken on dates years apart, van Maanen tied their swirling shapes to what he perceived as a real spinning motion. The spirals not only spun, van Maanen argued, but also their rapid rates of rotation suggested they resided within the Milky Way. At distances no greater than a few thousand light-years from the Sun, their spin pace remained within reason. If removed to the distance of an external galaxy, however, then the millimeters he had marked on the plates would translate into many times more kilometers traveled through space, and accelerate the rotation to exceed the speed of light. Since nothing could move faster than light, van Maanen's measurements of the spiral nebulae reduced the island universes to absurdity in Shapley's eyes.

"Congratulations on the nebulous results!" Shapley cheered to van Maanen on June 8. "Between us we have put a crimp in the island universes, it seems,—you by bringing the spirals in and I by pushing the Galaxy out. We are indeed clever, we are."

Shapley introduced himself to the wider Harvard community by offering a colloquium on astronomy, in which he tried hard to improve his performance over the previous year's "debate" in Washington. This time he told jokes. Former president Charles Eliot, who attended the talk, advised Shapley afterward that he need not embroider his grand subject with gratuitous humor.

In his attempt to win new friends for astronomy from Cambridge and Boston, Shapley instituted a series of open nights, inviting the public to hear a nontechnical lecture and peer through some of the telescopes. Admission was free, but interested visitors needed to register for tickets, since the observatory could not accommodate large crowds, and many sought entry. Pleased with the response, Shapley also planned to set aside separate nights for welcoming pupils from the local schools, as well as groups of boys' and girls' club members.

In the fall, when Hale inquired whether he should anticipate Shapley's return to Pasadena, Lowell said Harvard meant to keep him in the East.

University officials had voted to appoint Shapley as permanent director on the very day Hale's letter arrived, October 31, 1921.

As soon as Shapley relaxed into his leadership role, he awoke to the menace lurking in Mandeville. William Pickering released his latest research results in *Popular Astronomy,* and newspapers quickly picked up the Jamaica-based Harvard professor's account of "Life on the Moon." William reported vegetation sprouting on the lunar surface in rapid regular cycles, with abundant water and occasional steam emanating from craters. "We find, therefore," William asserted, speaking for himself, "a living world at our very doors, where life in some respects resembles that on Mars, but utterly unlike anything on our own planet, a world which the astronomical profession in general for the past fifty years has systematically neglected."

William was currently on sabbatical in Europe, a perquisite that Bailey had won for him from the Harvard Corporation. Bailey had tolerated William's excesses, and even gained him a small salary increase—the first raise in William's thirty-plus years on the observatory staff. "It seems to me that one can safely accept most of his observed phenomena," Bailey said in William's defense. "The difficulty comes in the interpretation." Shapley had no such patience. He planned to terminate Harvard's connection with the Woodlawn Observatory in Mandeville the moment William reached the age for mandatory retirement.

At the same time, but with a very different emotion, Shapley faced losing Miss Leavitt, whom he valued as "one of the most important women ever to touch astronomy." The discoverer of the period-luminosity law was dying of cancer. "One of the few decent things I have done," Shapley wrote in his memoir, "was to call on her on her death bed; it made life so much different, friends said, that the director came to see her."

Miss Cannon called often on Miss Leavitt toward the end, taking small gifts and marking each decline in her diary. "December 12. Rainy day pouring at night. Henrietta passed away at 10:30 p.m." On the fourteenth Miss Cannon attended "Henrietta's funeral at Chapel of 1st Cong. Church 2 p.m. Coffin covered with flowers."

• • •

SOLON BAILEY WITHDREW GRACEFULLY from the helm of the Harvard Observatory. In order to give the new director room to maneuver, Bailey offered to return to Peru for another sojourn at Arequipa. Along with his wife, Ruth, he anticipated a fruitful reunion with the southern star clusters. Their son, Irving, now a Harvard professor of botany and married to Margaret Harwood's sister Helen, did not accompany them this time. However, Miss Cannon did, encouraged by Shapley to take her own plates of the Milky Way, for later classification of stars fainter than ninth magnitude. She kept a close, lyrical account of her travels: "The sky line of New York faded away in a drapery of moist snow on March 1, 1922, when the Grace Liner *Santa Luisa* steamed out for Panama, Peru, and Chile."

It took two weeks for the ship to reach Mollendo, the nearest port to Arequipa, via the Panama Canal. Miss Cannon marveled at the Gatun and Miraflores locks, and even more at the sights above. "Epsilon and Iota Carinae, Kappa and Delta Velorum. How eagerly I gazed upon these stars, for my very first astronomical investigation concerned the spectra of bright southern stars never before visible to my eye."

Progress had altered the port of Mollendo since the perilous off-loading of the Bruce telescope in 1896. Miss Cannon was forced to bid the ship's company "a hasty farewell, for a novel method of landing passengers is customary at that port. A chair swung out from shore by a steam crane is let down into the tender, and seating himself, the passenger is quickly transferred to the Mollendo dock. After that, one expected novelties at every turn. And they came. Still 104 miles were we from Arequipa, with wonders all along the way." They traversed the desert and beheld the Andes. At the Arequipa railway station, a motorcar awaited to take them the final two miles to the observatory. "Such a drive through colorful Arequipa, the Cairo of South America, over the Chile River and into the town of Yanahuara, where streets are so narrow that pedestrians crowded against the walls of houses to avoid being run down."

The Arequipa station had been shuttered in November 1918, when the then supervisor, L. C. Blanchard, covered the telescope lenses and left to enlist in the armed forces. Even before the United States entered the war, decreased financial support had diminished the site's productivity, and the risks of shipping glass plates through war-troubled waters had grown prohibitive. Juan E. Muñiz, the longtime caretaker, watched over the locked, deserted station until peace reopened it. Frank E. Hinkley, a veteran of two previous Arequipa assistantships, took charge of the place in 1919, aided by the faithful Muñiz. Since Hinkley's departure in September 1921, Muñiz had single-handedly managed the building and equipment maintenance, the meteorological observations, and the taking of more than one thousand new pictures of the sky.

Observing in the transparent air at Arequipa revealed so much depth and detail that Miss Cannon had the sense she was gazing up into a live, long-exposure photograph. She learned to take her own plates with the various instruments, including "the unwieldy 24-inch Bruce Telescope. Each plate I secured was to me very precious. After developing and drying, I examined them as soon as possible, searching for new or unusual objects." One such object proved to be a new long-period variable; another, a nova.

"I expect to be an athlete when I return to Old Cambridge," she wrote Shapley, "for the running of the 13-inch requires turning a heavy dome, mounting ladders big and little, and all sorts of things, which Mr. Muñiz declared I could not do, for it was not 'woman's work.' I can do it all, however, except get good plates of faint spectra." A spry fifty-eight, she often walked the five miles to and from Arequipa "over the worst roads you ever saw" in the afternoon, and then worked five or more hours at the telescopes. "But it is great fun and does not tire me at all. Indeed it has been on clear nights so beautiful at mid-night that I hated to go to bed."

In addition to the pleasure of these pursuits, Miss Cannon enjoyed seeing a side of Bailey "in 'simpatica' Arequipa" that he rarely revealed in Cambridge. "A certain New England reserve and shyness," she noticed, "melted away under the tropical Peruvian sky."

Early in May, while the Baileys and Miss Cannon were thus occupied, most other astronomers reunited in Rome for the first general assembly of the new International Astronomical Union, the postwar incarnation of George Ellery Hale's old Solar Union. The meeting planned for 1916 had been scuttled by the conflict, but in 1919 scientists from many fields and twelve countries met in Brussels to forge new partnerships. The IAU emerged as one of the first of these forward-looking groups, and was personally welcomed into the world by King Albert of Belgium.

Although thousands of miles from the 1922 gathering, Miss Cannon was well represented in Rome. Henry Norris Russell, the chairman of the current Committee on Stellar Classification, had invited her participation early in 1919, and she had kept up a steady exchange of ideas with her fellow members since that time. Russell's formal report showed that Miss Cannon's system survived the discussions intact, strengthened by several additions useful to specialists in spectroscopy. For example, an S category had been introduced for a new class of red stars (thus lengthening the entrenched mnemonic to "Oh, Be A Fine Girl, Kiss Me Right Now, Sweetheart"). Also, the prefix of a lowercase *c*, a remnant from Miss Maury's more elaborate classification system, could now legitimately be placed before any capital letter to identify a star with markedly narrow, sharp lines. The *c* had proven its usefulness and won its rightful place in the stellar nomenclature. Similarly, the passage of ten years had underscored the importance of distinguishing between giant stars and dwarf stars, allowing the admission of *g* or *d* prefixes where appropriate.

Solon Bailey, as chairman of the IAU Committee on Variable Stars, had written the committee's report, but requested that Shapley read it for him at Rome. The report outlined a cooperative future: France, Italy, and other countries would henceforth coordinate observations, following the successful model of amateurs and professionals working together in the American Association of Variable Star Observers.

Harlow and Martha Shapley had vacillated through the early months of 1922 as to whether they could make the trip to Rome. Both their boys, Willis and Alan, fell dangerously ill that winter with pneumonia, and for a while the parents feared Willis would not pull through. When the crisis passed, Shapley still questioned the wisdom of a long absence from his new duties. Once he and Martha decided to go, however, he influenced the travel plans of other attending astronomers, and made sure to sail aboard the same ship as the Russells. He even convinced Arthur Stanley Eddington to move up the Royal Astronomical Society's centenary celebration from June to May, for the convenience of American foreign associates already abroad. Between the ending of the IAU meeting on May 10 and the start of the RAS events in London on the twenty-ninth, Shapley gave talks in the Netherlands on galactic structure and visited German observatories in Potsdam, Munich, Bergedorf, and Babelsberg.

In mid-June, seated once again at Pickering's revolving desk in the observatory, Shapley bragged to George Agassiz and the Visiting Committee of the trip's successes: "At the centenary of the RAS I talked about the work now being done at Harvard, and also made the principal address at a special meeting of the British Astronomical Association. At the Rome International meeting the Harvard Observatory astronomers were elected to 11 memberships on 8 of the 26 commissions, a recognition exceeded only by Mount Wilson among the American observatories, and my personal memberships naturally exceed those of any other American astronomer, because of the wide range of interest at Harvard, and were equaled only by those of the Astronomer Royal."

In other words, Agassiz should forget he ever doubted Shapley's competence to direct the Harvard College Observatory.

CHAPTER TWELVE

Miss Payne's Thesis

One might have expected Harlow Shapley to regret leaving the giant telescopes and ideal viewing conditions of Mount Wilson for life in a cloudy East Coast metropolis. Once settled in Cambridge, however, Shapley found he preferred his new role as observatory director to the rigors of making observations. "Observing was always very hard work for me," he conceded in his memoir. "I 'suffered' quite a bit those long cold nights. I suppose I didn't get as much sleep in the daytime as I needed, for I was running around observing ants in the bushes."

At Harvard he befriended his longtime correspondent, myrmecologist William Morton Wheeler, to whom he had mailed many vials of ants for expert identification. In the faculty dining club, Shapley, who had inherited Pickering's title as the Paine Professor of Practical Astronomy, hobnobbed with professors from other fields and honed his ideas about astronomy education. Although senior observatory staff members Solon Bailey, Edward King, and Willard Gerrish all went by the title of professor, they neither held doctoral degrees nor taught Harvard courses. The one man who did teach elementary astronomy at the university, Robert Wheeler Willson, did not associate himself with the observatory. Indeed, as Shapley pointed out, "the Observatory was not involved in instruction but in the production of knowledge." He determined to expand its mission to include the training of graduate students. Had a graduate astronomy program been in place at Harvard, Shapley said, President Lowell would not have needed to import "a Missourian fresh from California" as Pickering's successor.

Shapley well knew, from his own years as Henry Norris Russell's mentee at Princeton, that graduate students required graduate-student fellowships in order to survive. The only fellowship money on hand at the Harvard Observatory was the Edward C. Pickering Astronomical Fellowship for Women. Ergo, Shapley looked to the women's colleges as his sources of graduate students. In late January 1923, after much searching, he welcomed Adelaide Ames as his first recruit.

Miss Ames had attained graduation with honors, summa cum laude and Phi Beta Kappa, from Vassar College the previous June. An army officer's daughter, she had lived in the Philippines and traveled through China, India, Egypt, and Italy before attending high school in Washington, D.C. At Vassar, she had taken courses in integral calculus, molecular physics and heat, and physical optics and spectroscopy, which she documented by taping the catalogue descriptions into her Harvard application letter. She had also reported and edited for the *Vassar Miscellany News,* hoping to parlay that experience into a career. For months in the summer and fall of 1922, Miss Ames had tried but failed to land a job as a journalist. Now she was turning to astronomy as her second choice. Shapley had followed a similar path. Already a seasoned newspaperman when he entered college in 1907, he had chosen the University of Missouri on account of its much-touted new school of journalism. At enrollment, however, when he learned that the journalism school's opening had been delayed a year, he signed up instead for astronomy, physics, and classics courses. Despite his proficiency in Latin, he soon abandoned the classics for the activities and instruments of the campus observatory.

In the absence of a Pickering Fellow for 1921–1922, bank interest had accrued on the fund, enabling Shapley to increase the dollar amount of the award. He offered Miss Ames $650 to see her through two semesters of computation and research based on the glass plates, plus credit toward a Radcliffe graduate degree. He also agreed, now that she had made up her mind to become an astronomer, to let her start her academic year immediately, in the spring term, rather than make her wait until fall. When she arrived, he put her to work on one aspect of his pet problem—the distance and distribution of the stars in the Milky Way. Using plates from Arequipa, Miss Ames assessed and reassessed the apparent brightness of some two hundred southern stars. She also estimated their true—or "absolute"—magnitude from the intensity of selected lines in their spectra. Then she computed the differences between the apparent and absolute luminosity, with allowance for probable errors, to establish the stars' distances.

"Miss Ames, the new Pickering Fellow, is hard at it on Absolute Magnitudes," Miss Cannon noted in her diary on Monday, February 12. Later that week she reiterated, "Miss Ames, the new Fellow, is taking hold very well." In March, Miss Cannon mentioned in a letter to Caroline Furness at Vassar, the newcomer's former professor, that "Miss Ames is proving efficient and industrious and appears greatly interested in the problem of absolute magnitudes." By May a Harvard *Circular* reported the "Distances of Two Hundred and Thirty-three Southern Stars," under the joint authorship of Harlow Shapley and Adelaide Ames. With this new form of acknowledgment, Shapley out-Pickeringed Pickering. The late director had written virtually all the circulars himself, always giving credit to others in the text, but signing off at the end in his own name. Shapley made the researchers' bylines prominent on the first page, right under the announcement's title.

A new graduate student classmate for Miss Ames, Cecilia Helena Payne, came to Cambridge in the autumn of 1923, all the way from Cambridge, England. Miss Payne traced her interest in astronomy to the 1919 eclipse in Príncipe that had proved Einstein right. Although not a participant in the expedition, she had heard its leader, Arthur Stanley Eddington, lecture about it during her first year at Newnham, a women-only college of Cambridge University, where she was studying botany, physics, and chemistry. She experienced his talk as a "thunderclap," she said. He so inspired her that she returned to her dormitory room and wrote down his every word from memory, after which feat, and feeling her world transformed, she did not sleep for three nights. When she met the great Eddington during an open house at the university observatory, she avowed her desire to become an astronomer. He encouraged her by saying he could see "no *insuperable* objection." Other professors predicted at best an amateur status in the field for any Englishwoman, coupled with a paid position as a schoolmarm. Still Miss Payne persevered. She added courses in astronomy to her curriculum, studied the professional journals, learned how to compute orbits, reopened Newnham's long unused observatory, and with its small telescope began to explore the sky.

In 1922 a classmate took Miss Payne to London to hear Harlow Shapley address the Royal Astronomical Society. She already knew Shapley's name from papers he had written at Mount Wilson about globular clusters, but in person his youth and style surprised her. "He spoke with extraordinary directness," Miss Payne recorded, "conveyed the reality of the cosmic picture in masterly strokes. Here was a man who walked with the stars and spoke of them as familiar friends." Upon being introduced afterward to the speaker, she told him she wished to work for him in America, and Shapley humored her with his reply: "When Miss Cannon retires, you can succeed her." Surely he was joking, but Miss Payne seized the comment as reason to hope. She completed her college courses the following year and then, egged on by Shapley's promise of a Pickering Fellowship, she rounded up other prizes and grants to finance her move abroad.

Shapley stationed Miss Payne on the second floor of the Brick Building at Henrietta Leavitt's old desk. There Miss Payne, free at last from the Victorian strictures that had bound her since childhood, spent her new American-style independence on overwork. She arrived early at the observatory, stayed late, and sometimes failed to quit the place for days at a stretch. Rumor soon had it that the ghost of Miss Leavitt haunted the plate stacks and caused her lamp to burn through the night, but it was only Miss Payne, toiling till all hours.

"She is a healthy, but not really strong person," Miss Payne's widowed mother appealed to Shapley by letter from London, "and lives largely on her enthusiasms, and while I delight to think of her doing the work she loves, I cannot help being anxious at times lest she should not allow herself the necessary rest." Miss Payne's adoptive mother hens at Harvard, Annie

Cannon and Antonia Maury, shared Emma Pertz Payne's concerns, and vowed to protect her daughter. Nor were they alone in doting on Miss Payne. Professor Edward King, still the master of Harvard's photography, taught her the idiosyncrasies of the several telescopes. The night assistant, Frank Bowie, helped her develop her plates, and also informed her that the coordinates of any new comet—its right ascension and declination—might pay off handsomely if played in the local underworld's numbers game.

The tall, shy, ungainly Miss Payne and the lovely, engaging Miss Ames became fast, inseparable friends—also bridge partners at cards with Miss Cannon and her sister. Their closeness caused people to call the two students "the Heavenly twins." Between themselves, they affectionately referred to Shapley as "the Dear Director," or simply "the D.D." They liked the way he took the stairs two at a time, and the casual cheerfulness with which he heartened the underpaid female employees, often saying, "I think *I* could do this, so I'm sure *you* can." Miss Payne admitted to Miss Ames that she fairly worshipped the D.D.—that she might even be willing to die for him. Nevertheless, when Shapley suggested that Miss Payne carry on Miss Leavitt's work in photometry, she demurred. She preferred, she said, to pursue her own research agenda, applying the new theories of atomic structure and quantum physics to the analysis of stellar spectra.

No one at the Harvard Observatory had yet attempted such an investigation. No one possessed the background to undertake it. But Miss Payne hailed from Newnham College and the famed Cavendish Laboratory of Cambridge University, a place peopled with pioneers in these nascent fields. The Cavendish was home to Sir J. J. Thomson, recipient of the 1906 Nobel Prize in Physics for his discovery of the electron. Thomson's disciple Ernest Rutherford, whom Miss Payne described as "a towering blond giant with a booming voice," was the discoverer and first explorer of the atomic nucleus, and also the 1908 Nobel laureate in chemistry. During Miss Payne's student days at the Cavendish, she had learned the complex architecture of the "Bohr atom" directly from Niels Bohr, the 1922 Nobelist in physics. Although none of Bohr's lectures, which he delivered in a heavy Danish accent, ever lodged in Miss Payne's memory the way Eddington's relativity talk had stuck, she took good notes and saved them for later reference.

Shapley gave Miss Payne permission to do as she pleased, with unfettered access to the glass plate collection. Suddenly fearful of handling the precious materials, she worried aloud, "What if I should break one of the plates?" In that case, he assured her with his typical lightness, she could keep the pieces.

SOLON AND RUTH BAILEY HAD RETURNED to Peru in March 1922 with every expectation of remaining at the southern observatory for a period of several years. Their plans changed of necessity, however, when, within weeks of Miss Cannon's October departure, Mrs. Bailey suffered a stroke. It affected the left side of her brain, disturbing her speech and causing a partial paralysis on her right side. The dutiful Bailey tended to her with advice from local physicians and the help of a maidservant. He sent reports of Ruth's convalescence north, along with the plates he continued making of the Large Magellanic Cloud. Shapley, sympathetic to the couple's plight, sought to relieve them of responsibility at Arequipa, and looked to Edward King and his wife, Kate, as possible replacements. King, sixty-two years old, was Bailey's lifelong friend, eminently skilled and altogether willing, but doctors at the Harvard Medical School deemed him unfit for hard work at high altitude.

In March 1923 Margaret Harwood of the Nantucket observatory went to Arequipa to aid the Baileys. She took photographs through the Bruce telescope for Solon and also gave Ruth the benefit of her postwar experience with the Home Service of the American Red Cross. "I enjoy the work here very much," she wrote Shapley in June. "I now work with the Bruce the second half of the night. The dome works easily enough and so does the telescope. . . . This is a lovely spot. So far I have found only three kinds of ants, and they do not look very unlike New Englanders, but you may know better when you see the specimens."

By August Mrs. Bailey, still unable to speak or write clearly, was advised to go home, as the chances for her full recovery looked better at sea level. She went to stay with her son and daughter-in-law in Cambridge until her husband could join her.

Shapley's ongoing search for a new southern director led him to the Yerkes Observatory in Wisconsin, where Dorothy Block, the second Pickering Fellow, had found employment after leaving Harvard. At Yerkes, Miss Block fell in love with visiting astronomer John Stefanos Paraskevopoulos. After they married, she moved with him to his native Greece. The newlyweds were working together at the National Observatory of Athens when Shapley tapped them to take over the Arequipa station. Dr. and Mrs. "Paras" reached Peru in December 1923, freeing Bailey to go home at last. The circle of his longtime friends in the region bid him good-bye with the parting gift of an honorary doctor of science degree from the ancient University of San Agustin, along with an honorary title as professor of astronomy.

The energetic young Parases overcame the cloudy season in Arequipa by temporarily relocating the observatory. They carried two of the telescopes to a site near Chuquicamata in northern Chile, at an altitude of 9,000 feet. There they took large numbers of photographs under clear, dark skies, until April came and rendered Arequipa desirable again.

The shortened observing season in Arequipa made Shapley revisit Pickering's idea of transferring the Boyden Station to a new site in South Africa, but budgetary restraints prohibited such action. Shapley first had to meet the observatory's more pressing needs. Private donations from George Agassiz and other members of the Visiting Committee enabled the director to install an automatic sprinkler system in the Brick Building. Although Pickering had judged the building's brick exterior an ample safeguard against fire, Shapley feared its wooden floors, shelves, plate cases, desks, and other office furniture posed flammable threats to the treasured glass images.

"Since the first photograph of a star was made at Harvard in 1850 under the supervision of Professor George P. Bond," Shapley reminded President Lowell, "the Observatory has been a repository for an ever-growing collection of astronomical photographs." The building now housed nearly three hundred thousand glass plates. "The photographs made before 1900 are especially serviceable in the study of stellar motions and variability, and they are, of course, irreplaceable, and are unduplicated at other observatories." To the relief of uneasy astronomers, the new sprinklers finally gave the collection the protection it deserved. Tests of the system proved that water gushing from the sprinklers would do no damage to the glass universe, which was now further shielded inside new metal cabinets—dust-tight, mold-resistant, and moisture-proof.

What the observatory needed next were three or four more assistants to probe the plates. As soon as Miss Ames completed her fellowship year, in January 1924, Shapley hired her to fill Henrietta Leavitt's still-vacant place on the staff. No matter that Radcliffe College would not confer the master of science degree on Miss Ames until commencement day in June. She was ready now to help Shapley search the arms of the spiral nebulae for evidence of star formation. One thousand new spirals had just turned up in a single recent photograph taken by Bailey at Arequipa through the Bruce telescope.

Shapley pushed Miss Payne to go beyond the master's stage—to carry her original research all the way through to a Ph.D. Only a few other female astronomers had won this rare accolade, from universities in New York, California, and Paris; Miss Payne would be Harvard's first. Already she had drawn significant, publishable results from her studies. She was about to submit a report on the spectra of the hottest stars to *Nature*, under the name C. H. Payne, when Shapley challenged her by asking, "Are you ashamed of being a woman?" The question induced her to change her author identity to Cecilia H. Payne. Weeks later, when another of her finds struck Shapley as grist for another publication, he rushed her to prepare a paper right away, to be submitted in time for the next day's posting deadline. In his eagerness he even volunteered to type it for her. "What a glorious evening!" Miss Payne proclaimed her impromptu partnership with the D.D. "I wrote, he typed, far into the night. Into the mail it went. And I walked back to my room in the dormitory in a dream. My feet did not seem to touch the ground . . . it was almost like flying. I had not wanted to tell him that I was quite a good typist myself."

Miss Payne happened to be in Shapley's office the day he received a letter, dated February 19, 1924, from Edwin Hubble, a former colleague of his at Mount Wilson. "Dear Shapley," it began. "You will be interested to hear that I have found a Cepheid variable in the Andromeda Nebula." Few announcements could have rattled Shapley more than this one. The Andromeda Nebula, dimly visible to the naked eye, was the largest, most closely observed of all the spirals. A nova had erupted at its heart in August 1885, but was never captured on glass, given the primitive state of celestial photography at that early date. Since then the Andromeda Nebula had borne no evidence of individual stars, either at its center or anywhere along its spiral arms. Shapley's friend Adriaan van Maanen, who had measured Andromeda's rotation, swore he saw the nebula spinning rapidly, which meant it must lie relatively nearby—near enough for its individual stars to be seen if any such existed. Now Hubble, in a series of long exposures made on consecutive nights with the 100-inch telescope, had revealed whole congeries of stars in Andromeda.

"I have followed the nebula this season as closely as the weather permitted," Hubble's letter said, "and in the last five months have netted nine novae and two variables." The light curve he constructed for one of the variables showed the slow dip and rapid rise to maximum brightness characteristic of Miss Leavitt's Cepheid stars. Hubble's newfound Cepheid peaked near the very faint magnitude of 18, though its long period of thirty-one days decreed it must be thousands of times brighter than the Sun. The star appeared dim only by dint of great distance. Using Shapley's own calibration for the Cepheids, Hubble placed the Andromeda spiral more than a million light-years away. For the nebula to loom so large across a chasm that wide, it had to rival the Milky Way in size. Therefore the Andromeda Nebula must be a galaxy—an island universe—in its own right.

After Shapley read Hubble's news and looked at the light curve, he held out the pages to Miss Payne, saying, "Here is the letter that has destroyed my universe."

In a "Dear Hubble" reply of February 27, Shapley sounded unwilling to admit defeat just yet: "Your letter telling of the crop of novae and of the two variable stars in the direction of the Andromeda nebula is the most entertaining piece of literature I have seen for a long time." Rather than situate the variables *within* the nebula, as Hubble claimed, Shapley allowed only that they lay *in that general direction.*

Shapley had never much liked Hubble. Both men were Missouri-born, but Hubble, after spending three years as a Rhodes scholar at Oxford, had exchanged his midwestern twang for an affected British accent. He also clung to the military rank he had achieved in the army during the Great War, so that he continued to introduce himself as Major Hubble in civilian life. When, at Hale's invitation, the major reported to Mount Wilson in September 1919, he came dressed in jodhpurs and a cape. During the short period in which Shapley and Hubble observed on the same mountaintop, the one winced every time the other said "Bah Jove!" Even so, Shapley judged Hubble's meticulous work to be beyond reproach: "The distance of your variable from the nucleus and the lovely number of plates you have now on hand," Shapley conceded in his letter, "of course assures you of genuine variability for these stars."

Hubble had intended the news of the Cepheid for Shapley's eyes only, as he planned to confirm the Andromeda distance by further observations before making a public statement. The week after he dispatched his bombshell, however, Hubble took time off to marry Stanford University graduate Grace Burke Leib, a wealthy widow from Los Angeles, and honeymoon with her for three months in Europe. When he returned to work he turned up eleven more Cepheids in Andromeda. Shapley had once worried that eleven "miserable" Cepheids in the Milky Way lent insufficient support to his Big Galaxy. Now Hubble's dozen in Andromeda dealt the lone-galaxy idea a decisive blow. Hubble's Cepheids discredited van Maanen's measurements of rapid spiral rotations. Indeed, Hubble's Cepheids populated the cosmos with multiple "island universes."

"After Hubble's discovery of Cepheids," van Maanen wrote to Shapley, "I have been playing again with my motions and how I look at the measures." He continued to believe in them, though others ceased believing.

Heber Curtis, Shapley's former "debate" opponent, reveled in the proven reality of the island universes. Writing in *Scientia* in 1924, Curtis thrilled at the implications of the new insights: "Few greater concepts have ever been formed in the mind of thinking man than this one, namely,—that we, the microbic inhabitants of a minor satellite of one of the millions of suns which form our galaxy, may look out beyond its confines and behold other similar galaxies, tens of thousands of light-years in diameter, each composed, like ours, of a thousand million or more suns, and that, in so doing, we are penetrating the greater cosmos to distances of from half a million to a hundred million light years."

A human being might not be "such a big chicken," as Shapley had gibed in 1918 when he moved the Sun far away from the Milky Way's center, and yet, the human mind could traverse space and time.

CECILIA PAYNE PATIENTLY SIFTED the same objective-prism plates that had passed through the hands of Nettie Farrar, Williamina Fleming, Antonia Maury, and Annie Cannon. In the runic line patterns, which had helped her predecessors sort the stars into categories, Miss Payne read a new subtext. It concerned the actions of individual atoms, absorbing and releasing tiny quantities of light. Each spectrum's thousands of Fraunhofer lines registered the leaping of electrons from one energy level to another as they orbited atomic nuclei.

Miss Payne's vision was informed by the work of Indian physicist Meg Nad Saha of Calcutta, the first person to link the atom to the stars. In 1921 Saha demonstrated that the various classes of stars displayed their distinctive spectral patterns because they blazed at different temperatures. The hotter the star, the more readily the electrons around its atoms leapt to higher orbits. With enough heat, the outermost electrons broke free, leaving behind positively charged ions with altered spectral signatures. Saha created mathematical equations for predicting the locations of Fraunhofer lines in the spectra of various elements at extremely high temperatures—higher than could be achieved in laboratory furnaces. Then he fit his predictions to published spectra from the Harvard collection. The matches suggested that the categories of the Henry Draper classification depended almost entirely on temperature. The O stars were hotter than the B, which were hotter than the A, and so on, all the way down the sequence to its end.

Other investigators, from early classifier Angelo Secchi to current theorist Henry Norris Russell, had likewise noted the correlation between temperature and star type, but no one before Saha ever provided a physical mechanism for it. From the placement and intensity of certain Fraunhofer lines, Saha could estimate actual temperature ranges for stars in the various Draper categories.

Following Saha's promising lead, Edward Arthur Milne, one of Miss Payne's instructors at Cambridge, had reformulated and improved his techniques. Milne and his associate Ralph Fowler derived different stellar temperature values, though still in the order of the Harvard system. Fowler and Milne also factored in much lower pressures for stellar atmospheres than Saha had assumed, given that the gases around stars found ample room in which to spread thin. Compared with the air pressure at the surface of Earth, the minuscule atmospheric pressure of a star might be measured in the tiniest fractions of ounces per square inch, and these rarefied conditions would increase the tendency of atoms to ionize.

By 1923 Fowler and Milne had solidified the connection between atomic transitions and the intensity of corresponding Fraunhofer lines. A new research direction now lay open: By closely inspecting the strengths of the lines through the various spectral classes, a careful analyst might extract the comparative abundance of each component element. The raw material for making such revelations already existed in America, in the plate vaults in Cambridge and Pasadena. When Miss Payne departed Newnham College for the Harvard Observatory, Milne urged her to mine its glass photographs for spectra that would test and verify the Saha theory. "I followed Milne's advice," said Miss Payne, "and set out to make quantitative the qualitative information that was inherent in the Henry Draper system."

Princeton's Henry Norris Russell felt drawn to the same pursuit. Since Princeton lacked the needed resources, Russell arranged for extended leave periods on Mount Wilson, and sent one of his graduate students, Donald Menzel, to examine the plates at Harvard. Menzel's training in laboratory spectroscopy complemented the atomic knowledge Miss Payne brought to bear, but the two did not collaborate.

"I pressed on alone," Miss Payne wrote of her early struggle. "It was clear that some quantitative method must be devised for expressing the intensities of spectral lines, and I set up a crude system of eye estimates. Next came the identification of the line spectra, the selection of known lines for examination, and the arduous task of estimating their intensities on hundreds of spectra." Often she felt bewildered. She tasted her first breakthrough with the spectral lines of the element silicon, which she detected among the hottest stars, and in four successive stages of ionization (from neutral atom to the loss of one, then two, and finally three electrons). With these observations she calculated the heat required to remove the electrons, and thereby determined the temperatures of O stars to range from twenty-three thousand to twenty-eight thousand degrees.

Sometimes Miss Maury, who also liked to work late at night, stopped in with tidbits from her current spectral study of the southern binary stars. The women's pleasurable discussions were "painfully punctuated by insect bites," Miss Payne said, because Miss Maury insisted on keeping the windows open but could not bear to kill the mosquitoes.

Element by element, Miss Payne estimated, plotted, and calculated her way through the spectra to take the temperatures of the stars. All of her figures described the heat of the stellar atmospheres—the stars' visible, superficial layers that gave rise to their spectra. Temperatures deep inside the stars could only be conjectured. No one knew the processes by which stars generated their great power.

Shapley, intent on seeing Miss Payne garner Harvard's first doctoral degree in astronomy, organized a formal faculty committee to draft a preliminary written examination for her. She passed the test on June 10, 1924. As an official candidate for the Ph.D., Miss Payne attended summer astronomy meetings in New Hampshire and Ontario, wondering all the while where she would find the money to continue her study. The Pickering

Astronomical Fellowship expired after just one year, with no provision for renewal.

At the August gathering in Toronto of the British Association for the Advancement of Science, Miss Payne reencountered her first idol, Arthur Stanley Eddington, and also Edward Arthur Milne. They warned her that astronomy opportunities for women in England had not improved, and she should remain on this side of the Atlantic if at all possible. Fortunately, Miss Payne, by virtue of being a female college graduate, younger than thirty, who wished to study in the United States while still the citizen of a foreign country, met every qualification for the Rose Sidgwick Memorial Fellowship of the American Association of University Women. The $1,000 she won in September from this source secured her second year at Harvard.

Resuming her measurements of stellar temperatures, Miss Payne also teased out the relative proportions of elements in the various types of stars. Whereas her temperature determinations had been hard-won but satisfying in their agreement with previous ideas, the new figures for abundances alarmed her. Given that stars consisted of the same ingredients that constituted Earth's crust, most astronomers supposed that the recipe proportions must also agree. Common earthly substances, such as oxygen, silicon, and aluminum, were expected to prove just as prevalent among the stars. Miss Payne's calculations in fact revealed that kind of correspondence for almost every material, save for two notable exceptions, hydrogen and helium, the two lightest elements. Hydrogen abounded in the stellar atmospheres. Helium also proliferated, but hydrogen appeared to be about a million times more plentiful in the stars than on Earth. The profusion of hydrogen and helium made all other stellar components appear mere traces.

In December Shapley sent a draft of Miss Payne's strange report to Russell, the reigning expert on stellar composition. Russell praised her approach but balked at her results. "It is clearly impossible," he told her on January 14, 1925, "that hydrogen should be a million times more abundant than the metals."

Miss Payne had exercised great care in her methodology. Still, there was no gainsaying an authority of Russell's stature or experience. Obediently, she tempered her conclusions. When she submitted the write-up in February for publication in the *Proceedings of the National Academy of Sciences,* she pointed to her "improbably high" percentages of hydrogen and helium, and presumed them "almost certainly not real." In a field as young as the physical chemistry of stars, anomalous outcomes were no cause for shame. Rather, they indicated pockets of mystery for others to investigate and explain.

SOLON BAILEY OFFICIALLY RETIRED from the Harvard Observatory on February 1, 1925, though he did not stop working there. Now seventy, he went on discovering and studying variable stars in globular clusters, and, at Shapley's suggestion, started writing an authorized history of the observatory.

William Pickering also retired in 1925. Like Bailey, William, too, continued to practice astronomy, maintaining the Mandeville observatory at his own expense. He purchased a new telescope after Shapley forced him to relinquish the one he had long kept on loan in Jamaica—the 11-inch refractor donated to Harvard by Mrs. Draper in 1886. As soon as the instrument was back in Cambridge, Shapley rededicated it to stellar spectroscopy and photometry.

The year 1925 brought belated recognition for Henrietta Leavitt, from an admirer who did not yet know that she had died. "Honoured Miss Leavitt," began the letter of February 23 from Gösta Mittag-Leffler of the Royal Swedish Academy of Sciences. "What my friend and colleague Professor von Zeipel of Uppsala has told me about your admirable discovery of the empirical law touching the connection between magnitude and period-length for the S. Cephei–variables of the Little Magellan's Cloud, has impressed me so deeply that I feel seriously inclined to nominate you to the Nobel prize in physics for 1926, although I must confess that my knowledge of the matter is as yet rather incomplete." The writer, a ferocious advocate for the recognition of women in science, had agitated in 1889 to gain a full professorship at Stockholm University College for the Russian mathematician Sofia Kovalevskaya. In 1903 he successfully pressed the Nobel committee to include Madame Marie Curie in the physics prize being awarded to her husband, Pierre, and their countryman Henri Becquerel, the discoverer of radioactivity.

Shapley responded to Mittag-Leffler on March 9: "Miss Leavitt's work on the variable stars in the Magellanic Clouds, which led to the discovery of the relation between period and apparent magnitude, has afforded us a very powerful tool in measuring great stellar distances. To me personally it has also been of highest service, for it was my privilege to interpret the observation by Miss Leavitt, place it on a basis of absolute brightness, and, extending it to the variables of the globular clusters, use it in my measures of the Milky Way. Just recently in Hubble's measures of the distances of the spiral nebulae, he has been able to use the period-luminosity curve I founded on Miss Leavitt's work. Much of the time she was engaged at the Harvard Observatory, her efforts had to be devoted to the heavy routine of establishing standard magnitudes upon which later we can base our studies of the galactic system. If she had been free from those necessary chores, I feel sure that Miss Leavitt's scientific contributions would have been even more brilliant than they were." Shapley requested permission to share this gratifying tribute from Swedish scientists—confidentially, of course—with Miss Leavitt's mother and brother.

Miss Payne often congratulated herself for having avoided the routine work foisted on Miss Leavitt. The spring of 1925 found her in what she called "a kind of ecstasy" throughout the six weeks she spent writing her thesis. In it she described the novel nature of her procedures, laid out the stellar temperature scale she had calibrated, and summarized the abundance of the chemical elements in the stars. Based on her earlier exchanges with Henry Norris Russell, she repeated the caveat regarding the huge ratio of hydrogen and helium. Once again she dismissed their colossal abundance as "almost certainly not real."

Just as Pickering had instituted the Harvard *Circular*s in 1895 to announce Mrs. Fleming's discovery of her second new star, Nova Carinae,

Shapley inaugurated the Harvard Monographs in 1925 to showcase Miss Payne's thesis. Instead of tucking her opus into a volume of the *Annals*, where it would have been distributed to subscribing observatories and scientific institutions, Shapley published *Stellar Atmospheres* in a hardcover edition and offered it for sale at $2.50 per copy. He sent one as a gift to Russell, who shot back his thanks, claiming, "I have eaten it up since I got it yesterday." Russell declared Miss Payne's the best doctoral dissertation he had ever read, with the possible exception of Shapley's own thesis on the orbits of eclipsing binaries. "I am especially impressed," Russell said, "with the wide grasp of the subject, the clarity of the style, and the value of Miss Payne's own results."

The stunning takeaway from her work was the revelation that all stars closely resembled one another in makeup. The lettered categories in the Draper Catalogue signified differences in temperature, not variations in chemical composition. Henry Draper would have been amazed.

The hydrogen question, however, still begged for resolution. If the numerous, intense spectral lines of hydrogen did not signify a true abundance, then what did they represent? As the most salient features in many spectra, the hydrogen line patterns had guided the sorting of stars into categories. Spectral shapes had dominated the preparation of the Henry Draper classification, in much the same way as Pickering, in those early days, had assembled his recreational jigsaw puzzles. He always kept the hundreds of pieces facedown, rejecting any hints from the pictures, and fitting the whole together by shape alone. The new view of spectral lines, now imbued with atomic import, made the prominence of hydrogen lines seem incongruous. This new puzzle appealed to Russell, whose leisure pursuits included solving the crosswords printed in newspapers. Russell openly compared the analysis of a complex spectrum to "the solution of a glorified cross-word puzzle." He was getting at the nuances of spectral spelling and definition by spending more time at Mount Wilson, leading astronomers there to photograph particular stellar spectra for his study, and also cooperating with physicists at the National Bureau of Standards, benefiting from their laboratory spectra of individual elements.

In an editor's foreword to *Stellar Atmospheres,* Shapley reminded potential readers that the application of atomic analysis to astronomy was a field still in its infancy. Miss Payne's book, he said, showed the general state of the problem but stood open to revision and expansion in the immediate future. Meanwhile, he concluded with pride, "The book has been accepted as a thesis fulfilling the requirement for the degree of Doctor of Philosophy in Radcliffe College."

At the same time as Miss Payne earned her doctoral degree, Miss Cannon accumulated two more, honoris causa. Wellesley College wanted to award her a degree on May 29, 1925, the very day she was set to sail for England, so she booked passage on a different ship departing a few days later. "I assure you, dear President Pendleton," Miss Cannon said in her acceptance, "that to receive such an honor from my Alma Mater, where my first astronomical work was attempted, where Professor Whiting first led my thoughts toward the marvelous and newly developing subject of spectroscopy, will be the greatest incentive to continued effort and increased zeal in the ever-widening fields of my chosen science of astronomy." In 1921, in contrast, when the University of Groningen invited her to Holland to accept an honorary doctorate in mathematics and astronomy, she had found the timing inconvenient and asked to have the certificate mailed to her. Nor had she been lured away for even one day from the observatory in 1923, when the national League of Women Voters named her one of the "Twelve Greatest Women Living in America" (along with social worker Jane Addams, suffragette Carrie Chapman Catt, and novelist Edith Wharton).

Following the Wellesley ceremony, Miss Cannon embarked for England to attend the mid-July general assembly of the International Astronomical Union at Cambridge University. She traveled solo this time, as her sister, sixteen years her senior, was nearing eighty and unable to accompany her. While most of the conference delegates boarded in students' quarters, Miss Cannon occupied a room in the observatory residence as the special guest of Arthur Stanley Eddington and his sister, Winifred. During the IAU assembly, Harlow Shapley presented an illustrated talk about Miss Cannon's progress on the Henry Draper Extension. She went next to stay with her friends Herbert and Daisy Turner at Oxford, where she became the first woman in the history of the university to receive an honorary doctor of science degree. Moving on to Greenwich, she took part in Sir Frank and Lady Caroline Dyson's celebration marking the 250th anniversary of the founding of the Royal Observatory. The royals attended, too, and Miss Cannon saved a clipping that described the queen's dress as a soft shade of blue, something between hyacinth and hydrangea.

Miss Payne, who also attended the 1925 astronomy meetings, remained in England for the summer, at home with her mother and sister, Leonora, an aspiring architect. (Her archaeologist brother, Humfry, was away on a dig in Greece.) "I could wish to be back at work," she wrote Shapley at the end of July; "it took a visit to Cambridge to convince me finally that a return to America is a cause for jubilation rather than resignation, which is a piece of useful business done." In the fall she rejoined the Harvard Observatory as a postdoctoral fellow. She rented an apartment in Cambridge and registered as a lawful permanent resident of the United States, looking forward to full citizenship and voting rights. Suddenly she found herself strapped for cash. Her previous stipends had been paid at the start of each month, which led her to expect a continuation of the same payment schedule. But now, embarrassed, she realized she must wait for her paycheck till month's end. To make up the urgent deficit, she pawned her jewelry and her violin.

CHAPTER THIRTEEN

The Observatory Pinafore

Cecilia Payne discovered she loved having a home of her own. Settled in the new apartment, she indulged with pleasure in what she called "the feminine urges" to cook, sew, and entertain. "When one is spending several years in bringing a project to fruition," she explained, "there is great satisfaction in producing a masterpiece in the kitchen in a couple of hours."

Miss Payne had pictured herself "a rebel against the feminine role," before recognizing that her real rebellion "was against being thought, and treated, as inferior." She did not at all mind being treated as different—"Of course women are different from men. Their whole outlook and approach are testimony to it"—so long as none of her fellow scientists looked down on her on account of her sex. She faced scant risk of that at the Harvard Observatory, where Annie Jump Cannon could bake a batch of oatmeal cookies for a meeting of the Bond Astronomical Club and then lecture authoritatively to the assembled about her latest findings in spectroscopy.

Miss Cannon had lately moved with her elder sister, Mrs. Marshall, into a pleasant bungalow on Bond Street, just beyond the border of the observatory grounds. She called the place "Star Cottage," and it hummed with observatory social life. A motto inscribed in calligraphy in Miss Cannon's guest book expressed the philosophy, "Since Eve ate apples / Much depends on dinner." The book's pages preserved occasions such as "Observatory maidens for supper" (with the signatures of all sixteen invitees), "Edward Fleming to lunch," and "Tea out of doors. Col. & Mrs. Ames, Adelaide."

The observatory community's paragon of domesticity, however, was surely Martha Shapley. Like the earliest female influences—the wives, sisters, and daughters of the previous directors—Mrs. Shapley had been introduced to astronomy through family ties. Even so, her own skill as a mathematician predated her marriage and exceeded her husband's abilities in that field. Having helped Harlow with the calculations for his Princeton dissertation, Martha went on to write her own papers for the *Astrophysical Journal* about the orbits of eclipsing binary stars. In Pasadena, she and Harlow coauthored articles on Cepheids. After the move to Cambridge, despite the duties of caring for four children (she gave birth to the fourth, Lloyd, on June 2, 1923), Mrs. Shapley continued computing the orbital elements for eclipsing binaries. Although she received no salary, her name appeared on numerous Harvard *Circular*s and *Bulletin*s reporting her contributions. At the same time, she carried on the hospitality tradition of Lizzie Pickering, often inviting visiting scientists to board with the family in the director's residence adjoining the observatory. Harlow's convivial style of leadership required Martha to throw frequent parties at which staff members and distinguished guests mingled, played Ping-Pong and charades, and made music. She herself was such an accomplished classical pianist that no one minded when the sounds of her practicing reached the offices. In her role as director's wife, Mrs. Shapley became widely and affectionately known as "first lady of the Harvard College Observatory."

The core group of older computers, originally hired by Pickering, soldiered on in the Brick Building under Shapley's new management. Louisa Wells had joined the observatory in 1887, Florence Cushman in 1888, Evelyn Leland, Lillian Hodgdon, and Edith Gill in 1889, followed by Edith's sister, Mabel, in 1892, and Wellesley graduate Ida Woods in 1893.

Miss Cannon and Miss Maury, also part of the observatory's old guard, had originally stood out because of their college education in astronomy.

By 1925 they were flanked by easily a dozen female students, alumnae, and holders of advanced degrees. Margaret Harwood, for one, had gone out West to earn her master's years before Shapley bent the Pickering Fellowship to the purpose of graduate study. Adelaide Ames and Cecilia Payne were currently guiding the research of two master's candidates, Harvia Hastings Wilson from Vassar and Margaret Walton of Swarthmore College. The observatory also embraced a new rank of female guest researcher, in the person of Professor Priscilla Fairfield, Ph.D. Miss Fairfield had earned her doctorate in astronomy in 1921 at the University of California, Berkeley, and since then taught courses in "Celestial Mechanics" and "Measurement and Reduction of Photographic Plates" at Smith College in western Massachusetts. When she first came to work at Harvard in the summer of 1923, she asked for only a modest allowance toward local living expenses. In 1925, in addition to her summer stay, she was making the two-hundred-mile round trip between Northampton and Cambridge on any weekend when rain released her from supervisory duty at Smith's student observatory.

The appreciative Shapley won a $500 grant for Miss Fairfield from the Gould Fund of the National Academy of Sciences. "I suggest that you proceed at once in spending the money," he advised her on November 23, 1925. "May I suggest also that you try to arrange to spend it rather rapidly and efficiently, for I believe that with a little help from here, and a considerable success with the Gould Fund money, Smith College will support the work in a year or so." Miss Fairfield was comparing the spectra and proper motions of giant and dwarf stars belonging to Draper class M, in order to more clearly define the line distinctions between them. She used the Gould income to pay her student computing assistants thirty cents an hour. "It seems to me," Shapley added, "that we have the possibility now of building up a useful measuring or computing bureau at Smith, with two objects in view—to get the scientific work done and save one's soul; to make Smith College the girls' school where graduate astronomical work is done." In a handwritten postscript, he conceded that last wish a "funny statement from a Radcliffe professor!"

Shapley naturally sought to expand his graduate astronomy program to include men as well as women. At first, with only the Pickering Fellowship to offer, he could do nothing more than refer qualified male applicants to opportunities elsewhere. That situation changed in 1926, thanks to the generosity of Visiting Committee chairman George Agassiz. The new Agassiz Fellowship allowed the admission of Frank S. Hogg, from the University of Toronto, as the first seeker of a doctoral degree in astronomy at Harvard (as opposed to Radcliffe) College. Mr. Hogg's arrival coincided with that of the new Pickering Fellow, Helen B. Sawyer, of Mount Holyoke College. It soon became apparent that Mr. Hogg, who analyzed the spectra of comets, and Miss Sawyer, who studied star clusters, awakened a more than scientific interest in each other. Their courtship overturned a private joke that had long circulated around the observatory: *Why is the Brick Building like Heaven? Because there is neither marrying nor giving in marriage there.*

AFTER THREE ATTEMPTS to salvage the Arequipa enterprise by decamping to sites in Chile for the cloudy months, John and Dorothy Paraskevopoulos accepted a new Harvard mission. Shapley's strenuous campaigning on behalf of the observatory had brought in $200,000 from the Rockefeller Foundation's International Education Board and an equal amount from sources within the university—enough to transplant the Boyden Station at last from Peru to South Africa. In November 1926 the Parases started packing the equipment for the voyage east. They planned to use the Bruce as the premier telescope at Bloemfontein until a larger, more modern one replaced it. Construction of a 60-inch reflector, destined to be the biggest in the Southern Hemisphere, was already under way in Pittsburgh at the firm of J. W. Fecker.

Pickering had purchased a 60-inch reflecting telescope in 1904, hoping to improve his program of visual photometry. That instrument, however, built by British astronomer Andrew Ainslie Common, behaved poorly, and Pickering, after a few years of tinkering with it, set the 60-inch aside.

Shapley scavenged one of the relic's glass mirrors for resurrection in the new 60-inch.

The transfer and enlargement of southern operations looked to be the observatory's most outstanding material achievement in thirty years. Overseeing the project from afar sapped Shapley's energy, but did not derail his other activities. Together with Miss Sawyer, the current Pickering Fellow, he was developing a classification scheme for the hundred-odd globular clusters encircling the Milky Way. With Miss Ames, he was looking beyond the clusters to the more distant spirals, now recognized as external galaxies, or "island universes" outside the Milky Way, and taking a census of them. Shapley was also hosting foreign visitors attracted to Harvard by the cache of glass plates: one moment he was saying good-bye to Ejnar Hertzsprung, who spent seven months in residence between 1926 and 1927, and in the next moment welcoming the new guest researcher, Boris Gerasimovič from Russia. Simultaneously, as the observatory's spokesman and chief fund-raiser, Shapley kept up a heavy schedule of public talks and also a series of popular radio broadcasts, which he was collating and editing for publication—all while writing his own Harvard Monograph about star clusters.

The director's obvious exhaustion made George Agassiz uneasy. "You are that enviable and rara avis, the right man in the right place," Agassiz reminded Shapley on May 20, 1927. "Don't injure your machinery by running it at too high speed. The traveler who keeps well within his reserve power will make a longer and more fruitful journey than he who tries to crowd into each day more than it will hold. Delegate your authority, or if that is not possible, cut down the product. Don't burn yourself out, you are too valuable a man." Shapley promised he would arrange a family vacation toward summer's end.

In July the Parases reached South Africa's Orange Free State and sited the new permanent facility fourteen miles northeast of Bloemfontein, at Maselspoort. The altitude of the low hill, or kopje, where they settled was 4,500 feet, only about half the height of an Andes peak. But the seeing, which was everything, was better than at Arequipa. The Boyden Station born on "Mount Harvard" now commanded the view over "Harvard Kopje." The city of Bloemfontein, opening its arms to the new scientific center, extended water pipes to Maselspoort at government expense, along with power and telephone lines. Within a few weeks the Parases were able to resume their sky patrol of the southern latitudes.

Summarizing the highlights of the year in his September report to President Lowell, Shapley outlined more than forty projects in progress, without trying to describe the research in non-astronomer's terms. He declined to list the previous twelve months' worth of observatory publications by title and author, as had been the practice in past reports, on the grounds that there were too many of them; a full bibliography would take up too much space. Shortly after he submitted the report, the director and first lady of the observatory announced the birth of their fifth child, Carl Betz Shapley, on October 11, 1927.

In November Lydia Hinchman, beneficent founder of the Nantucket Maria Mitchell Association, endowed yet another special fellowship for women at the Harvard Observatory. "This gift comes just at the right time," Shapley said in his note of thanks. "The day before its arrival Miss Helen Sawyer, one of the Radcliffe graduate students at the observatory, talked over with me the possibility of going on in her study for a doctor's degree in astronomy." Miss Sawyer, who had entered Mount Holyoke as a chemistry major, switched to astronomy in her junior year under the influence of Professor Anne Sewell Young. One event in particular had swayed her: "For the total solar eclipse of January 24, 1925, Miss Young was able to get a special train to take all the college people to a golf links in Connecticut, inside the path of totality. There the glory of the spectacle seems to have tied me to astronomy for life, despite my horribly cold feet as we stood almost knee deep in snow."

While still at Mount Holyoke, Miss Sawyer had developed "a particular fondness for globular clusters as my favorite celestial objects." At Harvard she worked with the world's authority on these objects, even entering her observations in the same record book that Shapley used. Miss Sawyer was also pleased to meet Solon Bailey, the clusters' first champion, and to immerse herself in the very plates he had taken with the Bruce telescope in Peru. Using these and other images, she helped Shapley divide the stellar conglomerations into several subclasses according to the concentration of stars at their centers. The differences hinted at various evolutionary stages of cluster development. Under Shapley's guidance Miss Sawyer also redetermined the apparent photographic magnitudes of all the clusters, in an effort to confirm the distance of each.

Miss Sawyer was poised to receive her master's degree from Radcliffe in June 1928, when Mr. Hogg would be granted his from Harvard. Together they looked forward to still higher academic distinction, and to much else besides. Frank thought he could complete the doctoral degree requirements in just one more year, but Helen needed at least another two, possibly three. Radcliffe insisted that she master the German language, and Mount Holyoke had given her insufficient grounding in mathematics and atomic physics. Indeed, when she attended her first Harvard colloquium to hear Miss Payne speak "On the Lifetime of an Excited Hydrogen Atom," she innocently mistook the title of the talk for an attempt at humor.

Miss Payne made the transition from postdoctoral fellow to full-fledged astronomer in 1927, drawing a salary from the observatory of $175 a month. As part of her new duties, she took over the editing of all in-house publications—annals, bulletins, circulars, and monographs. She edited with relish, professing a real love for "the hurly-burly of the University Printing Office, the look and smell of the fresh galleys, the detail of proofreading, the craftsmanship of makeup." She also drew diagrams for the other authors, and, in the case of foreign authors, rewrote their papers to improve their English if necessary.

Although not a professor, Miss Payne instructed the graduate students and supervised Frank Hogg's doctoral research. Shapley thought she deserved an academic position, and said so to President Lowell, who opposed the idea. Just as Lowell had once denied Miss Cannon a Harvard Corporation title as curator of the glass plates, he now refused to appoint Miss

Payne to the university faculty. Moreover, Shapley reported, the president swore that Miss Payne should never ascend to a Harvard professorship while he was still alive.

Miss Payne, though stymied in one direction, advanced along the avenues open to her. She took up Miss Leavitt's work with photographic magnitudes, as Shapley had asked of her early on. Also at Shapley's request, she began writing a new monograph—a follow-up to her *Stellar Atmospheres*—concerning stars of high luminosity.

As Miss Payne pored over the spectra of the most luminous stars, she wondered whether their light might have been dimmed on its passage through space. Perhaps an unidentified absorbing medium, either fine dust or opaque gas, stole some of the luster from starlight. If so, then the bright O stars must be even brighter than they appeared, and therefore closer than imagined. Distance alone would dim them down by known proportions, as Newton's inverse square law dictated: For two stars emitting equal light, the one twice as far from the observer shone one-quarter as brightly. Adding dust to the equation would make the farther star seem even farther removed.

Others before Miss Payne had weighed the likelihood of "interstellar absorption" of light. The observatory's own Edward King suspected a dimming effect, and had conducted photographic experiments over the years to detect it. Although King could not quantify the amount of light lost to space travel, he believed that some loss did occur. Shapley, on the other hand, argued confidently that no invisible interference dulled the glow of stars or spirals. In Shapley's view, only the obvious patches of obscuring matter, such as the dark lanes contained within certain star clouds, could swallow light. Outside those zones, he was sure no obstruction occurred. All Shapley's assessments of the distances to globular clusters, and of the Sun's distance from the center of the Milky Way, assumed light's unimpeded path through interstellar space. Miss Payne thought it best not to contradict the D.D. on this point. In deference to him, she adopted his opinion, and also adduced evidence to support it.

Shapley set as his goal the mapping of the whole Milky Way. He had made a good beginning when he removed the Sun from the galactic center, but that was only the first step toward a full picture. If one could see it from the outside, would the home galaxy assume a spiral form, with starry arms swirling about a bright central bulge? Or would it resemble one of the many blob-shaped, non-spiral galaxies? Or reveal a layout even more irregular? Shapley was counting on Cepheids and other variable stars to serve as waypoints on his quest. To that end he was seeking new assistants to identify new variables from the latest plates, and to track their changing magnitudes with Miss Payne's new photometric standards. Charting the Milky Way from the inside out, Miss Payne opined, was a bit like limning all of London and its environs from a city street corner, through a heavy fog.

"JUST WHAT ARE YOUR PLANS NOW," Shapley inquired of Priscilla Fairfield on May 26, 1928, "for the summer and for measuring proper motions of cluster type Cepheids and for spending some of your Gould Fund on a measuring girl at the Harvard Observatory?" He begged a quick reply, as he was en route to Europe for the general assembly of the International Astronomical Union, to be held in Leiden, and also the Heidelberg meeting of the Astronomische Gesellschaft: "I am leaving this side of the planet a week from tomorrow for two months."

Miss Fairfield wrote back on May 29 to say, "I have changed my plans and am going to the other side of the planet myself this summer. I hope this will only postpone and not prevent my measuring proper motions of cluster type Cepheids as I plan to return early in September."

July in Leiden saw the largest and most global gathering of astronomers ever united. For the first time the attendees, 243 in all, received star-shaped name badges to help them identify one another. New countries had signed on to the Union since its 1925 meeting, including Argentina, Egypt, and Romania. A new postwar spirit of rapprochement prevailed, enabling fourteen astronomers from Germany to participate freely in all discussions and activities—though they could not vote on policy, pending their government's adherence to the Union. Willem de Sitter, president of the IAU and director of the Leiden Observatory, had personally invited them. De Sitter's opening remarks recognized "the great German nation," defining the country's greatness in terms of "the number and importance of its contributions to astronomy."

The moment Miss Fairfield stepped off the train at the Leiden station, she attracted the attention of Dutch astronomy student Bartholomeus Jan Bok. He had been drafted by the local organizing committee as an official greeter of foreign delegates, particularly unaccompanied ladies such as the blond, petite Miss Fairfield and her Smith College superior, Harriet Bigelow. The genuine glow of his welcome only increased over the course of the week-long meeting. Everywhere Priscilla turned, "Bart" turned up beside her.

In addition to Shapley and Miss Fairfield, the Harvard Observatory contingent at Leiden comprised Cecilia Payne, Margaret Harwood, Antonia Maury, and Adelaide Ames. At this, her first general assembly, Miss Ames was elected to IAU membership. Also, in recognition of her studies of spiral galaxies, she was duly appointed to the IAU Commission on Nebulae and Clusters. Shapley wrote to her parents in Massachusetts, to let them know how much she seemed to be enjoying herself.

Miss Fairfield tried to fend off the amorous advances of her new suitor, who, at twenty-two, was a good ten years her junior. Bart Bok persisted, however, and at length overcame her misgivings.

Bok made a different but equally favorable impression on Shapley. Having studied under Willem de Sitter and Ejnar Hertzsprung at Leiden, he had devoured Shapley's papers about the Milky Way. Now pursuing a doctoral degree at the University of Groningen, Bok let Shapley know how earnestly he wished to work with him. Shapley, receptive as ever to the excitement of a young astronomer, thought that sounded like a fine idea.

In the formal IAU sessions at Leiden, the world's astronomers declared themselves still content with the Draper classification of the stars. Continued investigations had only confirmed its lasting practical value.

The widely admired author of that system, meanwhile, remained at home in Cambridge. Miss Cannon tended to her increasingly infirm elder sister and examined additional faint stars for a new installment of the on-going Henry Draper Extension. Helen Sawyer, who occupied the office next door to Miss Cannon's, could hear her calling out the letter categories "day after day" to her recorder, Margaret Walton. Miss Cannon pronounced the classifications about as fast as Miss Walton could write them down. Jesse Greenstein, one of the Harvard underclassmen just starting his astronomy studies, once remarked that while the average person might judge from a distance whether a particular animal was an elephant or a bear, "Miss Cannon could separate the rogue from the good elephant, or the grizzly from the brown bear, at a glance." Henry Norris Russell, on one of his regular visits to the observatory, suggested querying the aging Miss Cannon on her technique, but Miss Payne said it would be fruitless. She doubted Miss Cannon could explain her process, or even knew herself how she managed it. Her uncanny powers of instant recognition obeyed no conscious train of thought. She simply saw each star for what it was.

Russell, on the other hand, relied more on logic. In the years since convincing Miss Payne to call her findings "almost certainly not real," he had puzzled long and hard over the question of the hydrogen abundance. He gathered new data of his own during stays at Mount Wilson. More than once he concluded a series of calculations that suggested a predominance of hydrogen in the Sun and other stars. But each time this happened he rejected the results as spurious—until he could reject them no longer, and admitted the inescapable omnipresence of hydrogen. In a lengthy paper "On the Composition of the Sun's Atmosphere," which ran in the *Astrophysical Journal* in July 1929, Russell finally agreed with Miss Payne and cited her 1925 study. He made no mention of his earlier disbelief when he conceded, at the end of fifty pages, that "the great abundance of *H* [hydrogen] can hardly be doubted."

The constitution of the universe had inverted. The huge superabundance of hydrogen and helium, first intuited by Cecilia Payne, reduced all other cosmic components to chaff. What had long been presumed minimal was now proved plentiful by Russell's deep analysis: the lightest, most immaterial elements reigned supreme.

• • •

"YOUR KIND OFFER OF THE Agassiz Research Fellowship made me very happy and I accept it with both hands," Bart Bok wrote to Harlow Shapley on April 22, 1929. "Priscilla was delighted when she heard about the chance at Harvard. When it hadn't come, she had promised to come to Groningen, but now that we have this wonderful opportunity, everything looks so much finer. I'll never forget that you gave me a chance to work for the woman I love and I really shall do all I can, not to disappoint you."

The new graduate student arrived in the States on Saturday, September 7, married his betrothed on Monday at her brother's house in Troy, New York, and a week later wrote to Shapley to say he was enjoying "a most wonderful and happy time" on honeymoon in the Berkshires.

The stock market crash of October 1929 exerted no immediate ill effects on the observatory, where the mood remained expansive through December. Shapley, having proposed Cambridge as the venue for the semiannual meeting of the American Astronomical Society, invited the approximately one hundred members into the director's residence for a New Year's Eve party. The night's entertainment cast the staff in a fully costumed and staged performance of *The Observatory Pinafore.* Book and lyrics for the operetta had been written fifty years earlier, in 1879, by former telescope assistant Winslow Upton, using music from the 1878 Gilbert and Sullivan hit about a Royal Navy ship named for a ladies' apron. Upton apparently drew inspiration from the chorus of sisters, cousins, and aunts who boarded *H.M.S. Pinafore,* and he turned them into a bevy of female computers for his purposes. In place of the two boy babies switched at birth, Upton twisted a zany new intrigue around a pair of prisms stolen from one of Pickering's photometers.

Upton's libretto spoofed everything and everyone in the observatory. Since his stint there had overlapped Williamina Fleming's clerkship prior to her son's birth, he made sure to cite "our Scotch maid," who "has unfortunately returned to her native land."

An early scene finds young Upton, who lived in a garret off the stairway to the Great Refractor, ruing the noisy activities that wake him in his off-hours. Arthur Searle hears him out and suggests he find a room off-site. Upton vows, "I shall when my salary is large enough," to which Searle retorts, "I guess you'll die of old age in that room if you wait for a large salary before giving it up."

For all that had changed at the observatory since Upton's day, the wages of astronomy remained an evergreen theme. The point was driven further home in the lyrics: "An astronomer is a sorry soul, / As free as a caged bird; / His sympathetic ear should be always quick to hear / The directorial word. / He must open the dome and turn the wheel, / And watch the stars with untiring zeal, / He must toil at night though cold it be, / And he never should expect a decent salaree."

The current crew enjoyed reviving all the ghosts of New Years' past, especially the gentlemanly Pickering, whose strongest oath when provoked to anger in the drama is, "Oh Polaris!"

At an encore performance on Monday, January 13, for the monthly meeting of the Bond Astronomical Club, half again as many guests squeezed into the director's residence. "After that night," Shapley pledged with mock solemnity, "we resume our sober and methodical attempts to maintain the scientific standing of the Observatory."

Foreigners now looked on Harvard as "the meeting place of the astronomers of the world." They deemed its multinational character unusual, even in the decidedly international science of astronomy. Shapley was happy to receive Svein Rosseland from Norway in 1930, and Ernst Öpik from Estonia, even though, as he half-joked to George Agassiz, it was difficult "finding places for our present staff and scientific visitors to sit down." The workforce had about trebled in size over the course of his directorship.

A cloud darkened Shapley's horizon in the form of interstellar obscuration. In the spring of 1930, Robert J. Trumpler of the Lick Observatory in California produced proof that the Milky Way teemed with dust. Trumpler, who had once collected ants for Shapley in Australia, now showed that invisible particulates permeated the galaxy. The intruding dust undermined almost all determinations of magnitude, and likewise the distances deduced from magnitudes. Trumpler came to these conclusions by observing one hundred open clusters—close associations of stars that were not as densely crowded together as in globular clusters. He calculated each open cluster's distance two ways, by its apparent brightness and also its apparent diameter. Both values predictably decreased with distance, but the open clusters faded in brightness much faster than they shrank in size. Some form of "dark matter" definitely intervened to absorb their light. As far as Trumpler could tell, the mysterious absorbing medium was confined to the Milky Way, but unevenly distributed; it concentrated along the plane of the galaxy and dissipated near the poles.

Shapley had banked on transparency when he estimated the galaxy's extent at three hundred thousand light-years. With the effect of interstellar absorption factored in, the Milky Way contracted to about half that size. "This was a sharp rebuff to Shapley's ideas," Miss Sawyer observed, "and he felt it deeply." Still he wanted to inform the observatory community of the news. Miss Sawyer said he asked her "to review Trumpler's paper for a colloquium, knowing, I think, that my empathy for the situation would lead me to deal with it as unabrasively as possible." That presentation was her last at Harvard. In September she and Frank Hogg got married at her family home in Lowell, Massachusetts, with most of the observatory family as witnesses. The couple set up housekeeping in South Hadley, near their new workplaces—she as assistant to Professor Anne Sewell Young at Mount Holyoke (with time off to write her dissertation on globular clusters), and he as a researcher assigned to the 18-inch telescope at Amherst College.

Margaret Walton, the Swarthmore graduate and former Pickering Fellow who now served as Miss Cannon's recorder, married R. Newton Mayall, a landscape architect and amateur variable star observer. She had met him while assisting Margaret Harwood one summer on Nantucket. Miss Walton, however, retained her position at the Harvard Observatory, and also the use of her maiden name.

Marriage no longer spelled the end of a career for a woman astronomer, as it had when Miss Cannon entered the field. She approved of the new trend, and defended the rights of all her Pickering Fellow brides: "Does it not appear that research, which is not confined to fixed hours or necessarily to office walls, may easily be carried on by married women?" she asked rhetorically in one of her regular reports as chairman of Nantucket's Astronomical Fellowship Committee. "A stellar photograph may be studied at home, during odd hours, and perhaps may not require more time from a wife or mother than is frequently given to bridge playing or various other social activities."

Solon Bailey spent six years writing *The History and Work of Harvard Observatory, 1839 to 1927.* The completed volume was published as a Harvard Monograph early in 1931. Only a few months later, on June 5, Bailey died of a short, sudden illness at his summer home in Norwell. His wife and son were with him. The bereaved Miss Cannon borrowed a line from *Julius Caesar* for her friend's obituary: "His life was gentle." Having known Bailey for thirty years, she could honestly report in the *Publications of the Astronomical Society of the Pacific,* "He won the respect of all by his wide sympathy, his justice, his never-failing kindness, and his complete lack of self-seeking."

Bailey's other good friend and colleague, Edward King, wrote a separate obituary for publication in *Popular Astronomy.* Before that story had a chance to appear, however, King himself fell ill and died on September 10, just ten days after he retired. Miss Payne, who had grown close to King through their shared passion for collecting old classical texts, wrote a tribute to him in a subsequent issue of *Popular Astronomy.* She quoted a letter that King had received from Bailey the previous spring, when both men were reminiscing over their long careers in astronomy: "To have done work which is widely recognized, to have gained the sincere esteem of many and the real love of even a few, surely these are sufficient reasons to look on life as well worth the living."

Despite the Depression, Shapley added more than $1 million in gifts and bequests to the observatory's endowment in 1931, most of it from the

Rockefeller Foundation. Construction began in July on a bigger brick building, adjacent to the old one, with all the latest innovations in fireproof protections, plus room for the plate collection's projected growth over the next fifty years. In October Shapley announced that several of the photographic telescopes would soon move from Observatory Hill to a secluded spot in the woods northwest of Cambridge, near the village of Harvard, Massachusetts. Stands of maple, oak, pine, and birch at the new site, called Oak Ridge, would shelter the instruments from wind, soot, and the plague of artificial light. Shapley also made public his plan to install a 60-inch reflector—a northern counterpart to the one being built for Bloemfontein—at Oak Ridge. The sylvan thirty acres would soon boast the best-equipped observing station in the eastern United States.

Shapley's "contributions to astronomical science" had recently won him the National Academy of Sciences medal perpetuating the name of Dr. Henry Draper. The small, elite fraternity of Draper Medal winners included Edward Pickering, George Ellery Hale, Henry Norris Russell, and Arthur Stanley Eddington. Shapley thought the time ripe to add a woman's name to that roster, and he submitted his nomination in support of Miss Cannon.

"Her life's work under the auspices of the Memorial founded by the founder of the medal is drawing to a close," Shapley wrote to the members of the academy's Draper Fund Committee. "It is needless to comment on the nature and permanence of her contribution." She was the one who had put the classification system into its current, popular form, and she alone had passed judgment on each of its quarter million stars. "The Henry Draper Catalogue, as far as I remember, has never received any official recognition in the United States by medal, vote, honorary degree, or otherwise," Shapley continued. "Miss Cannon is quite indifferent to such recognition; but it strikes me that the work she has done is the greatest contribution to science enabled by the Drapers and that there is a certain appropriateness in considering Miss Cannon for the medal."

The committee concurred. Shapley, in his glee, prepared a private citation in advance of the official one yet to come. His read as follows:

Dr. Annie Jump Cannon

the benign presence of the Brick Building, noted collector of degrees and medals, author of nine immortal volumes and several thousand oatmeal cookies, Virginia Reeler, bridge player, and the godmother of SW Andromedae, and especially the recipient of the

Draper Medal of the National Academy of Sciences

the first medal ever bestowed on a woman by that honorable body of fossils and one of the highest honors attainable by astronomers of any sex, race, religious or political preference—in recognition of this great honor indicated by the Draper Medal and on behalf of the staff of the Harvard Observatory I anoint you with the usual star dust—present you with the metagalactic emblem of good luck—and drape you with this token of becoming the world's jolliest Draper Medalist.

CHAPTER FOURTEEN

Miss Cannon's Prize

The upcoming festivities at the observatory promised to be Harlow and Martha Shapley's biggest party ever: the triennial general assembly of the International Astronomical Union, scheduled for September 1932. When he invited the IAU to Harvard, the director begged to let four years instead of the usual three—just this once—elapse between meetings. The delay would enable an entertainment comparable to that at past receptions in the capitals of Europe, where potentates of church and state had presided over opulent ceremonies. In place of those trappings, the more homespun Harvard general assembly promised the natural miracle of a total solar eclipse. Shortly before four o'clock local time on Wednesday afternoon, August 31, 1932, the Moon would obliterate the Sun, and the skies over New England would go dark. The visiting astronomers could distribute themselves along the path of totality from Quebec through parts of Maine, Vermont, New Hampshire, and Massachusetts, then pack up their equipment and repair to Cambridge.

Shapley's plan entailed a big risk of bad weather. The available predictions gave only a fifty-fifty chance of ideal observing conditions on eclipse day, and failed eclipse expeditions would surely pitch the astronomers into gloom. Even Miss Cannon had turned sour (briefly) when cloud cover cheated her out of photographing the flash spectrum during the 1923 eclipse in southern California. Nevertheless, the director banked on success. He put the first graduate student, Adelaide Ames, now his sidekick in galactic exploration, in charge of hospitality arrangements for the IAU.

In May 1932, as the time of the assembly neared, Shapley and Miss Ames were wrapping up their extensive, definitive survey of external galaxies. They had looked at more than a thousand island universes contained in the Harvard plate collection. They had classified the shapes of these nebulae, seven hundred of which were spirals, and computed the total brightness of each according to a uniform system of photometric magnitudes. The Shapley-Ames Catalogue showed, for the first time, the distribution of such objects over the entire sky. Although Shapley's picture of the Milky Way still lacked detail, he and Miss Ames had taken steps toward tracking the contours of the larger cosmos. It was larger than they imagined, and apparently growing larger all the time. As early as 1914, Vesto Melvin Slipher of the Lowell Observatory in Arizona had shown the spectra of most spiral nebulae to be shifted toward the red, meaning they were receding along the line of sight—rushing away at high speed. Later Edwin Hubble at Mount Wilson built on Slipher's findings. By painstakingly approximating the distances to the fleeing spirals, he perceived a new relationship, dubbed Hubble's law: the farther the galaxy, the faster it fled.

In late June, after Miss Ames had left the text and tables of the Shapley-Ames Catalogue with the college printer, she set off with a few of her observatory coworkers for a short vacation at Squam Lake, near Holderness, New Hampshire. The family of her friend Mary Allen owned a rustic lakeside camp there with panoramic views of the White Mountains. On Sunday the twenty-sixth, Miss Ames and Miss Allen paddled a canoe out to the center of the lake, where a squall struck and capsized them. They laughed over the mishap at first, while trying to right the canoe. Then they abandoned ship and swam for shore. Both young women were strong swimmers, but when Mary, in the lead, reached the shallows and looked back over her shoulder, there was no one behind her. She called out to Adelaide a few times. She screamed for help. Others rushed to the scene and made repeated dives at the spot where the hysterical survivor had last seen her companion's head above water. They could find no sign of her. Someone had to telephone Colonel Thales L. Ames, commanding officer of the Springfield Armory, to inform him that his thirty-two-year-old daughter had apparently suffered a cramp while swimming, and drowned.

When the news reached Shapley on Monday, he closed the observatory and drove to the lake with Leon Campbell to assist Colonel Ames. Police were there, directing search parties on foot and in small craft. Airplanes skimmed over the lake in circles until dark. On Tuesday Miss Cannon wrote in her diary, "Adelaide's body not found yet." It took more than a week to recover her remains. During the funeral service at Christ Church, Cambridge, on July 7, Miss Cannon said it broke her heart to look at Colonel and Mrs. Ames. The following day they buried their daughter—their only child—in Arlington National Cemetery.

THE TRADITIONAL BARRING OF WOMEN FROM CAREERS in most fields of science led, in 1897, to the founding of a small organization centered in Boston called the Association to Aid Scientific Research by Women. In its early years, the group's sole function was to raise funds in support of a research table for American ladies at the Zoological Station in Naples, Italy, where Professor Anton Dohrn extended a Pickering-esque welcome to female researchers. Within a few years the association broadened its mandate to award grants to individual scientists, and later a prize, the Ellen Richards Research Prize, recognizing lifetimes of accomplishment. The award honored the late Ellen Swallow Richards, a chemist and charter member of the association who had been the first woman admitted as a full-time student to MIT. She established the Women's Laboratory at the institute in 1876, and, after several years of teaching with neither salary nor title, she became assistant professor of chemical analysis, industrial chemistry, mineralogy, and applied biology. Even then she was given no compensation, but, as a married woman—the wife of Robert Hallowell Richards, head of mining engineering at MIT—she could afford to work without pay.

The Association to Aid Scientific Research by Women awarded the 1932 Ellen Richards Research Prize of $1,000 to two worthy recipients, Dr. Helen

Dean King, a biologist at the Wistar Institute of the University of Pennsylvania, and Dr. Annie Jump Cannon of the Harvard College Observatory. With that decision, the twelve members declared themselves satisfied with the progress they had seen, and they drafted a resolution to dissolve the organization. "Whereas," it said, "the objects for which this association has worked for thirty-five years have been achieved, since women are given opportunities in Scientific Research on an equality with men, and to gain recognition for their achievements, be it Resolved, that this association cease to exist after the adjournment of this meeting."

An outside observer might have judged the association's dissolution premature. Miss Cannon seemed to think so, as she moved to extend its good works. "Your letter concerning the Ellen Richards Research Prize is of great interest," she replied to Marjorie Nicolson, dean of students at Smith College, on June 10, 1932, "and the enclosed check is certainly a delightful recognition of my many years of astronomical investigation." She felt doubly grateful, she said, because she had known the prize's namesake, and recalled several conversations with Mrs. Richards—at the College Club of Boston and at meetings of the Association of Collegiate Alumnae—concerning opportunities for women.

"I wish I might convey through you to the Committee," Miss Cannon continued, "to the donors and to all the members of the former Association to Aid Scientific Research by Women, my very great appreciation of this prize. I hope to use it to advance, in some way, astronomical research by women. Moreover, the very thought of it will be a constant spur for increased efforts on my part towards the completion of the various problems which I have under way, realizing that the faith of so many representative women demands a justification in the highest service which it is possible for one to give."

Miss Cannon earmarked her $1,000 bounty to endow the Annie Jump Cannon Prize. She wished it to be awarded biennially or triennially by the American Astronomical Society, to a deserving woman of any nationality. It would take time, she knew, for interest on the seed money to grow into a substantial purse, but she did not want to put off the first award indefinitely. She was nearly seventy years old, and determined to confer the Cannon Prize in person at least once while still in good health. She thought she might enhance the cash value with a feminine token of some kind—a brooch or a necklace in a starry motif, which could be kept as a memento and worn long after the money had been spent. She began searching for a craftswoman who could fabricate the item she had in mind.

NOT ONLY ASTRONOMERS FLOCKED to the Northeast for the total eclipse of August 31, 1932. Wide publicity turned the event into a popular tourist attraction, and although most scientists who made observations were looking at the sky, a few measured the eclipse's effect on terrestrial phenomena such as radio transmissions and animal behavior. William Morton Wheeler, the Harvard myrmecologist, had read historical reports of ants that ceased all activity during eclipses, as though transfixed by the sudden midday darkness. Wheeler felt certain the ants were reacting to the rapid drop in temperature, not the absence of light. Eager to learn more, he put out a call for voluntary reports from interested parties.

"Field observation on insect behavior is, of course, difficult during a total eclipse," Wheeler conceded in the *Proceedings of the American Academy of Arts and Sciences,* "because the observer is apt to be very desirous of witnessing at the same time a wonderful astronomical event, which he may never again be able to contemplate." Still, Wheeler hoped that entomologists would forgo gaping skyward to look down and amass basic information that could be tested further during future eclipses. "Even the astronomers," he said, "had to learn what to expect during a total eclipse before they could make elaborate preparation beforehand."

True to form, astronomers geared up for the occasion. A full boatload of instrumentation from the Royal Observatory, including a 45-foot-long telescope, sailed from Greenwich on July 13, to ensure the accompanying scientists ample time for assembly and practice on-site. Those who planned on just *seeing* the eclipse, as opposed to *observing* it, could afford to arrive at the last moment.

In August a consortium of Canadian astronomers descended on Louiseville, Quebec, to array their instruments on the local fairgrounds, where they encountered the French expedition and the eclipse party of the American Amateur Astronomical Association of New York. All three groups enjoyed optimal conditions and met most of their observing goals during the 101 seconds of totality. Only twenty-five miles to the north at Saint-Alexis-des-Monts, as luck would have it, an equipment-laden encampment that had been set up weeks in advance suffered an invasion of clouds and could not move out from under them. Similarly, a group in Gorham, New Hampshire, reported complete failure due to weather, although four members of that team managed to snatch a clear view of totality through a gap in the clouds by driving thirty miles eastward in a fast car. The other partners stayed put and carried out their fully rehearsed program regardless, on the off chance that the clouds might part in mid-eclipse, though that did not happen. Overall, only a fortunate few of the researchers won the eclipse-day gamble. These included the several Harvard parties, especially the principal one stationed at West Gray, Maine. At a spot near West Acton, Massachusetts, someone noticed a nuptial flight of ants emerging from the soil to mate in midair—a behavior known to be cued by falling temperature.

Many of the nearly two hundred delegates to the Cambridge IAU meeting swapped eclipse stories as they gathered on Saturday, September 3, for the start of ceremonies at Alice Longfellow Hall in the Radcliffe Yard. Bernice V. Brown, the dean of Radcliffe College, could not resist dropping an eclipse allusion in her opening address: "We are used to welcoming students to the College who come here looking at the world through rose-colored glasses," she said, "but this is the first time we have had visitors with smoked glasses."

The students were all absent on summer holiday, freeing the lecture halls and dormitories for the astronomers' use. Dean Brown said she hoped some of the guests would return another time when classes were in session. "The Harvard Observatory," she acknowledged, "has not only been ready and willing to give instruction to the Radcliffe girls, but they have fostered a long line of graduate students. We are delighted to show hospitality to their colleagues."

Sir Frank Dyson, astronomer royal of Great Britain and current president of the IAU, thanked Dean Brown and saluted Charles Francis Adams, the secretary of the navy, who conveyed official greetings from President Herbert Hoover. Turning to Shapley, Sir Frank recalled an earlier visit: "In 1910 on the way to the Solar Union at Mount Wilson we saw the many activities in which Professor Pickering was engaged. We are all delighted to come again. We are glad to see the pleasant face of Miss Cannon once more. We are all delighted to see the Harvard Observatory and all the activities in which you are engaged, especially your researches on the Milky Way."

Addressing the full audience, Sir Frank welcomed one and all, with special reference to the representatives from Germany and Austria, whose countries were not yet officially connected to the Union. Then he asked the assembly to stand as he recited the names of the twenty-two IAU members who had died since the Leiden meeting. "Some of these," Sir Frank said, "like Monsieur Bigourdan, Father Hagen, and Doctor Knobel, have died in the fullness of years, but others, like Monsieur Andoyer, General Ferrié, and Professor Turner, we hoped to have with us for many more years, and we can ill spare them. We particularly wish to express our sympathy to Professor Shapley and the staff of the Harvard Observatory on the tragic death of their gifted and charming colleague, Miss Adelaide Ames, who was the secretary of the local committee charged with the preparations for this meeting. We are mindful of the services to astronomy of all those we have lost and we shall keep them in affectionate memory."

THE DEATH OF MISS AMES devastated her "Heavenly twin," Cecilia Payne. Miss Payne had been among the invited guests at Squam Lake, and could not bear to speak of what had happened there. One day she bumped into an acquaintance who, having heard a bungled account of the accident, blurted out, "Why, I heard you had been drowned!" To which Miss Payne confessed that she wished it were true, and that the lake had claimed her life instead of Miss Ames.

In the aftermath, Miss Payne compared herself—"absorbed in my work, shy and unattractive"—with the other twin, whom she idolized as beautiful, outgoing, and beloved by everyone. Miss Payne resolved to try giving more of herself in the future—to "embrace life and do my part as a human being." Open and vulnerable, she fell in love for the first time, "unreasonably, groundlessly, but nonetheless thoroughly (for I am nothing if not thorough)," she wrote in her memoir. "It did not take me long to see that my love was not, never would be reciprocated, and I fell into a state of despair." Priscilla and Bart Bok buoyed her through the dark hours, encouraging her to go away for a while. She took their advice and planned an extended trip to visit the observatories of northern Europe.

In the summer of 1933 Miss Payne traveled to Leiden, Copenhagen, Lund, Stockholm, Helsinki, and the historic Tartu Observatory, with its 9-inch refractor built by Joseph von Fraunhofer, in Estonia. Everywhere she went she was welcomed and indulged. At the IAU meeting in Cambridge the previous summer, she had renewed her friendship with Boris Gerasimovič, who invited her to visit him at Pulkovo. Once in Europe, however, other hosts tried to persuade her to skip the Russian leg of her journey, as did the American consul in Estonia. The United States had no diplomatic relations with the Soviet Union, the official told her, and would not be able to help her out of any difficulties she might encounter there. But Miss Payne did not heed these entreaties; she meant to honor her promise to Gerasimovič.

Miss Payne found herself alone on the train after it crossed the Russian border. At Leningrad, Gerasimovič came to meet her with a driver in a pickup truck, but, since it was illegal for three to ride in the front seat, she had to sit on the floor of the truck bed all the way to the observatory. "I spent two weeks at Pulkovo," she recorded, "and felt I had experienced a lifetime. The atmosphere of tension never lifted. It was not only the drab and squalid living conditions of the man who was the Director of one of the great Observatories. It was not only the scarcity of food—for food was severely rationed and they shared their rations with me. I had brought them some coffee, and they gave a party to celebrate it—nobody there had tasted coffee for several years. One day there was a special treat for supper—carrots, and my host confessed that he had stolen them from a neighboring garden.

Small wonder that the food nearly choked me—unappetizing as it was, I was taking it from their plates. Everyone was afraid—afraid to talk lest they should be overheard. One of the young women . . . led me to the middle of a wide field and begged me in a whisper to help her to go abroad—'I would wash dishes,' she said, 'I would do *anything* to get away from here.' And what could I do? What could I possibly have done? I was appalled."

Miss Payne lost all sense of personal grief in those grim surroundings. She felt as though she were holding her breath for the duration of her stay, and she carried that sense of oppression away with her on the train ride to Germany. In Göttingen in August, still shaken, she attended the meeting of the Astronomische Gesellschaft held at the Mathematics Institute. She spotted her mentor Eddington there, but dared not try to attach herself to his distinguished circle. Instead she took a seat by herself in the rear of the large auditorium. A young stranger of about her own age sat down near her, asking in German, "Are you Miss Payne?" He introduced himself as Sergei Illarionovich Gaposchkin. He had ridden a bicycle from Potsdam to the meeting, one hundred fifty miles, in hopes of finding her, he said, and he gave her an autobiographical sketch that explained his desperate situation. She read it that night, in lieu of sleep. Gaposchkin was a Russian émigré facing Nazi persecution. As one of ten children born to poor parents in the Crimean village of Yevpatoria, he had worked on fishing boats, farms, and in factories to realize his childhood dream of becoming an astronomer. He had studied in Bulgaria and Berlin, and written a doctoral dissertation about eclipsing binary stars, in which he cited papers by both Harlow Shapley and Cecilia Payne. Just recently he had lost his job at the Babelsberg Observatory for political reasons. Gaposchkin was under suspicion in Germany of being a Soviet spy, and had been denied reentry into Russia, where authorities presumed him a German spy. "Of course I knew I must help him to escape the last of the many disasters that had overtaken him," Miss Payne realized. "When I saw him the next day I told him that I could make no promises, but I would do what I could."

Gaposchkin's first sight of Miss Payne surprised him, he later wrote, because he had expected her to be as old as the famed Harvard astronomer

Annie Jump Cannon. Her youth and bearing put him in mind of "a ripe peach left alone on a tree, darkened, wrinkled a little outside, but the more delicious inside."

Miss Payne found it relatively easy to convince Shapley of the need to rescue Gaposchkin. Since the early 1920s, the director had been engaged in various efforts to aid Russian astronomers affected by war, revolution, and civil strife. Yes, Shapley said, they would make room for Gaposchkin at Harvard, but how to get him there? He was stateless and destitute. Miss Payne, who had become a naturalized United States citizen in 1931, went to Washington to expedite the granting of a visa to a man without a country.

On Sunday, November 26, 1933, Gaposchkin sailed into Boston Harbor, and Miss Payne met his ship at the pier. She drove the new arrival to the apartment she had found for him in Cambridge, then took him to meet Shapley and the rest of the observatory staff that same evening at a party in the director's residence. Given that Gaposchkin spoke very little English, Miss Payne continued to converse with him in German. They had almost constant occasion to speak to each other, as he was assigned to work under her direct supervision on the new photometric standards. Even his one-year salary of $800 came out of the funding for her project. Their familiarity bred affection. After three months of working together, they eloped to New York and were married in City Hall on March 5, 1934. Shapley, informed in advance of their plans, facilitated the wedding through New York friends, who not only served as witnesses at the ceremony but also treated the couple to a nuptial luncheon of champagne and caviar. The next day the bride wrote to Shapley from the Hotel Woodstock, "I had never thought that such happiness could be for me."

Shapley broke the news of the Payne-Gaposchkin marriage to the observatory community during one of his "Hollow Squares." These informal meetings took place weekly in the library of the newly expanded Brick Building. They got their name from the temporary rearrangement of the reading tables into a rectangle, with the chairs ranged around the outside so that all participants could face one another. Shapley used the Hollow

Squares (renamed "Harlow Squares" by the graduate students) to share research developments at other observatories, introduce visiting astronomers, and give the Harvard staff a forum for reporting their own progress or suggesting new ideas.

Apparently no one had noticed the romance blossoming between Miss Payne and her Russian research assistant, for the general reaction was one of shock, even outrage: why, aside from being two lonely astronomers in their midthirties, the pair had absolutely nothing in common, and what's more Cecilia, at five feet ten inches, stood half a head taller than her new husband, who had *no* prospects and precious little to offer her, so far as anyone could tell.

THE PASSAGE OF TIME embellished the story of the wedding announcement at the Hollow Square. People liked to say that Miss Cannon reacted by fainting dead away, but of course she did nothing of the kind. The union of two scientists, she knew, could produce a whole greater than the sum of its components. Single or married, Cecilia was still her top pick for the first Annie Jump Cannon Prize, to be awarded at the December 1934 meeting of the American Astronomical Society (AAS), in Philadelphia. It just so happened that the current president and first vice president of the society were men particularly important to the winner's career, Henry Norris Russell and Harlow Shapley.

The interest earned on Miss Cannon's $1,000 principal amounted to an inaugural prize of only $50. However, she had located an able jeweler, Marjorie Blackman, to fashion the desired gold pin in the form of a spiral nebula. After a few tries in silver, Miss Blackman, who was new to astronomy, mastered the nebular proportions, smoothed an area on the back for engraving, and attached a loop that allowed the pin to be worn also as a pendant on a chain. Miss Cannon was pleased. "I think it is quite pretty," she wrote to AAS secretary Raymond Smith Dugan shortly before the meeting. "Isn't it the first universe ever made by a woman?"

At the banquet on December 28, Russell presented the award to Mrs.

Gaposchkin for "her valuable work in interpreting stellar spectra." She gave a brief acceptance speech, and then invited Miss Cannon to make a few anecdotal remarks on the preparation of the Henry Draper Catalogue.

More and more these days, Miss Cannon was called upon to share her memories. Her recorder, Margaret Walton, had taken to typing up some of the anecdotes and filing them in folders with labels such as "Under Southern Stars" and "Old Dover Days." Miss Cannon remembered some details from childhood as clearly as she recalled her stellar classifications. "In the house where I was born," she dictated to Miss Walton, "there stood on the white marble mantel, a candelabra representing a gilded tree. At the base two children are about to waken a sleeping huntsman. Five outspreading branches support the candles, which are surrounded by glass prismatic pendants. I remember no earlier plaything than these prisms which were easily detachable. To hold one in my hands, to catch a sunbeam, and watch the brilliant prismatic colors dance over the wall was a delight to my youthful eyes. Even now I hold one of these pendants in my hands, and note that it is embossed with stars. Stars and prisms! How prophetic was this baby amusement of the profession which was destined to fill my life."

Miss Cannon continued to classify ever fainter stars with gusto, but publication lagged on account of budget shortfalls. In 1937 Shapley solved the problem by changing the format. In place of the tabular rows and columns of digits typical of the Henry Draper Catalogue and the early installments of the Henry Draper Extension, the new "Henry Draper Charts" were presented as reproductions of photographic plates. On these plates, Miss Cannon numbered the spectra and designated the class letter for each, and sometimes assessed the magnitude as well. Every annotated picture thus compressed several hundred stars' worth of data. Assistants no longer needed to list the individual star alongside its other catalogue designations, or describe its position in declination and right ascension. These shortcuts saved so many pages of type that five to ten times the number of spectra could be published annually. There was no need for Miss Cannon to slow down.

Along with her classifying, Miss Cannon also kept up her bibliography pertaining to variable star observations. The fifteen thousand index cards she inherited in 1900 had since multiplied many times over, and now numbered around two hundred thousand. She maintained as well a much smaller collection of astronomical verse—poems by Milton, Longfellow, Tennyson, and others—within the covers of a slim notebook. She liked these lines from Ralph Waldo Emerson's "Nature" well enough to transcribe them: "Teach me your mood, O patient stars! / Who climb each night the ancient sky, / Leaving on space no shade, no scars, / No trace of age, no fear to die."

Now in her seventies, Miss Cannon still reported to the observatory six days a week. Every spring she selected a new Pickering Fellow and a new recipient for financial aid from Nantucket nonagenarian Lydia Hinchman. The fresh faces of the incoming young ladies had gradually replaced the long-familiar ones. Florence Cushman retired in 1937 after forty-nine years of computing, proofreading, and assisting Willard Gerrish, who soon followed her. Lillian Hodgdon, the assistant curator of the photographic library, left at the end of half a century's service. Miss Hodgdon's title, like Miss Cannon's, was an observatory honorific, not a university position. In January 1938, however, five years after James B. Conant succeeded Abbott Lawrence Lowell as president, the Harvard Corporation officially recognized Miss Cannon as the William Cranch Bond Astronomer and Curator of Astronomical Photographs. In the same stroke, the corporation reversed the discrimination against Mrs. Gaposchkin by naming her the Phillips Astronomer.

"G O S H !" observatory secretary Arville Walker exclaimed in an in-house broadside announcing the twin events. "For the first time in its 301 glorious years, the Corporation of Harvard University has deliberately recognized women academically" by these appointments. "It is an occasion. Let's celebrate with a Dutch-treat luncheon—Commander Hotel—Tuesday, January 18, 12:30—85¢. Please report your plans promptly to Miss Walker." Fully fifty well-wishers showed up.

In her excitement, Miss Walker had slightly overstated the nature of the honors done to Miss Cannon and Mrs. Gaposchkin. While their new titles had indeed been granted by the corporation and approved by the Board of Overseers, they were not exactly academic. Miss Cannon was to continue as before, in a post now glorified by association with the name of founding observatory director William Cranch Bond. The Phillips name, as in "Phillips Astronomer," had also been attached to the institution since its infancy. Edward Bromfield Phillips, a Harvard classmate and close friend of Bond's son George, had willed his family fortune of $100,000 to the observatory shortly before he killed himself in 1848, at the age of twenty-three. William Cranch Bond consequently became the first Phillips Professor, followed by his successors, George Phillips Bond and Joseph Winlock. In Pickering's day, the bequest of an even larger fortune changed the director's title to Paine Professor, in memory of benefactor Robert Treat Paine. At that point, the Phillips Professorship devolved on Arthur Searle, and passed to Solon Bailey after Searle officially retired in 1912. Now that the Phillips title belonged to Mrs. Gaposchkin, it gained her a listing in the Harvard catalogue as an officer of the university. Shapley hoped it would do yet more. In pressing for her appointment as the Phillips Astronomer, he had needed to assure the corporation that conferring the title on a woman would not make her a member of the college faculty or even of the astronomy department. Meanwhile Shapley confided to President Conant, "At some future time, if the University approves the policy, I should like to recommend that the title be changed to Phillips Professor of Astronomy." After all, she was already teaching and directing graduate research, sitting on three IAU commissions, and enjoying an international reputation as an astrophysicist, spectroscopist, and photometrist. She was also the mother of two children. Edward, named for her father, was born May 29, 1935, and Katherine on January 25, 1937. The Gaposchkins had bought a house in Lexington on a large lot, where they cleared the yard of rocks and brambles to make room for flowers and trees.

The Annie Jump Cannon Prize, awarded every three years by the executive council of the American Astronomical Society, increased gradually in cash value over time. In 1937 it went to Charlotte Moore Sitterly, who was Henry Norris Russell's personal computer. With each new winner, Miss Cannon gave a new craftswoman the chance to create her own decorative universe in jewelry. The 1940 prize recognized Julie Vinter Hansen of the Copenhagen University's Østervold Observatory, an expert on calculating orbits for comets and asteroids. Although Miss Vinter Hansen was working in the United States at the time of the award presentation, she could not leave Berkeley to attend the banquet, being held once again in Philadelphia. As soon as the check and accompanying token reached her by mail in January 1941, she wrote to thank Miss Cannon: "The 'medal' has now arrived and to my pleasant surprise it is no medal at all. I love this feminine touch that it has taken the form of a pin that can be worn every day. I think it is very beautiful and it has been on my dress ever since it came, also yesterday when I was radio-interviewed in Oakland and had an occasion to tell how grateful I am to this country and its astronomers, an occasion that was so much more welcome to me, as I missed the opportunity to express what is in my heart by not being able to go to Philadelphia."

In the next breath Miss Vinter Hansen asked, "Why do you not come out to California and its wonderful climate this winter? I know the astronomers would be delighted to welcome you." Miss Cannon declined, too taken up with activities to consider the trip. "Last Saturday I gave a broadcast by short wave on 'The Story of Starlight,'" she told her Oxford pen pal Daisy Turner, the widow of Herbert Hall Turner, on January 21. "Dr. Shapley was sick in bed with flu, and heard it horizontally, he said. He spoke a good word for it, which pleased me greatly. There is something uncanny about speaking to such a hypothetical audience. . . . Do you recall the Bond Club? It still goes on, and I am heading a course in astronomical reading for a group meeting me in a fortnight. . . . I am just as busy as a bee with many interesting things outside. My dear neighbor Ruth Munn was just in talking about the Cambridge Historical Society meeting to be held in her home next week. She wants me to wear my best evening gown."

The 1941 IAU general assembly, originally planned for Zurich in August, had already been canceled owing to the alarming escalation of aggression in Europe. "Oh I do hope Oxford will not be harmed," Miss Cannon worried to Mrs. Turner. "It is all so merciless, horrible and unbelievable." Not to dwell on war, Miss Cannon turned to other news of mutual friends and mundane events. "We are having cold weather, but crystal clear, brilliant sunshine, invigorating air. It is bracing, and I feel 'fit.'" She closed with "Love again and again, A.J.C."

She continued to work and feel well until mid-March, when her health took a turn. After a few weeks the illness grew serious enough to send her to the Cambridge Hospital, where she died on Easter Sunday.

"On the thirteenth of April, 1941," Cecilia Payne-Gaposchkin reported in *Science,* "the world lost a great scientist and a great woman, astronomy lost a distinguished contributor and countless human beings lost a beloved friend by the death of Miss Annie J. Cannon."

Cecilia could remember the day, not so long ago, when she and Sergei invited the whole staff out to Lexington, to what would have been a garden party if not for the torrents of rain, and how Annie had breezed in, full of cheer, wearing a bright, flowered dress, and hoping that her outfit might "do something to counteract the weather." Although Miss Cannon was well into her eighth decade at the end of her life, it could still be said that she had died young.

"During the past year," Shapley rued in his 1941 annual report, "the observatory suffered a heavy loss through the death of Miss Annie Jump Cannon. In her seventy-seventh year Miss Cannon was still engaged in classifying the spectra of the stars, work in which she was a pioneer and which she had carried on for more than forty years. During that interval the spectra of about half a million stars passed under her scrutiny. In addition to the results published in the Henry Draper Catalogue and the Henry Draper Extension, she had classified approximately one hundred thousand stars, unpublished at the time of her death.

"To commemorate the life and work of this gentlewoman, whose kindly advice, enthusiasm, and perseverance charmed and encouraged all who met her, the Observatory has planned a series of memorials. A memorial volume of the Annals, to contain the one hundred thousand unpublished spectra, will soon be issued. The expense of this volume has already been met through the generosity of her friend Professor James R. Jewett, *Emeritus* Professor of Arabic in Harvard University. Two endowed fellowships in the Observatory are contemplated as a further memorial; they would continue to offer the inspiration of Miss Cannon's example to young men and women interested in undertaking the career of astronomical research. One would be available to students from Wellesley College, Miss Cannon's alma mater, and would be awarded with preference to students from Delaware, her native state. The office in which Miss Cannon worked has been set aside as a Memorial Room and will soon be redecorated in an appropriate manner. The type of work that she did will be carried on in this room and in the others that she used."

CHAPTER FIFTEEN

The Lifetimes of Stars

Cecilia and Sergei Gaposchkin had the observatory almost to themselves throughout the war years of the 1940s. Often they brought their children to work, letting them sled down the steep slope of Observatory Hill or play hide-and-seek in the dusty catacomb under the Great Refractor. The couple had a third child now, Peter, born April 5, 1940. They also owned, in addition to their house on Shade Street in Lexington, a small farm near Townsend, where a neighbor helped them raise pigs and poultry for local markets. As naturalized U.S. citizens, they considered their farm labor a patriotic duty, and they delivered their meat and eggs by horse and buggy to conserve their allotment of rationed gasoline. In 1942, when Japanese Americans on the West Coast were forced to enter internment camps, the Gaposchkins took in the family of Reverend Casper Horikoshi, whose son and daughter were playmates of Edward and Katherine.

To increase their own and others' understanding of the global crisis, the Gaposchkins set up a discussion group called the Forum for International Problems. It met one evening every other week in the observatory library, with Shapley's hearty approval. Speakers came from all departments of the university as well as the Boston and Cambridge communities. In her role as chairman, Mrs. Gaposchkin tried not to take sides, particularly, she said, when participants "urged their arguments with intemperate zeal." At times she feared the disputes on the platform would lead to physical violence.

From Cambridge to Oak Ridge to Bloemfontein, young men dispersed to the branches of the armed forces. Staff members were stunned, Shapley noted, to discover how well their training and practice in astronomy had fitted them for "effective cooperation in the war effort." After all, sailors needed to steer by the stars, and an astronomer's lenses, mirrors, and photographic techniques lent themselves readily to strategic purposes. The director informed President Conant in autumn 1942 that twenty-five of his subordinates were engaged in eleven distinct types of military research, some of which were too confidential for comment. Mrs. Shapley was working for the navy, calculating ballistics trajectories. Frances Wright, one of the younger generation of computers, devoted herself practically full-time to teaching celestial navigation, as did Bart Bok. Shapley was often called away on matters pertaining to refugee scholars, such as "committee meetings in New York to raise money to rescue people from Hitler's grasp." President Conant assumed the chairmanship of the new National Defense Research Council, with responsibility for projects pertaining to nuclear fission, and disappeared occasionally from his Harvard office to visit undisclosed locations in the Midwest and Southwest.

Although half the telescopes at Oak Ridge shut down for lack of graduate students to operate them, the Boyden Station in Bloemfontein remained at peak activity. The new 60-inch reflector saw constant use, as did the old 8-inch Bache, the 13-inch Boyden, and the big Bruce photographic telescope. July and August 1942 brought a spell of unusually fine winter weather to South Africa, allowing the Parases to break all their previous records of work accomplished. They had to stockpile most of the plates on-site, however, until ocean transport again became safe.

With no war-related assignments, the Gaposchkins stuck to their studies of variable stars. From the twenty thousand variables discovered over the preceding fifty years of nightly all-sky photography, they chose two thousand that had shone brighter than tenth magnitude at least once. Then they followed these target stars through plates dating back to 1899, and established the light curve for each one, to classify the type of variation it displayed. They also checked back on some "new stars" to see what had become of them since their eruptions as novae. The star called U Scorpii, for example, which first called attention to itself as Nova Scorpii in 1863, had flared again, they found, in 1906 and 1936—outbursts that made U Scorpii the first known "recurrent nova." The plate collection had kept the 1906 news a secret for decades; the 1936 event had also gone unnoticed till now. Like an oracle, the great glass storehouse teemed with knowledge, but divulged it only when petitioners posed a specific question.

In their long collaboration, husband and wife divided the kingdom of the variables roughly in half. Cecilia specialized in the Cepheids and other "intrinsic variables," which alternately brightened and dimmed on their own, while Sergei looked at stars that repeatedly hid some or all of their light behind a partner. He had a "nose" for picking out surprising star pairs. For example, he showed that the giant star VV Cephei not only varied its light in typical Cepheid fashion, but also was partially eclipsed every twenty years by a small companion. No one else had noticed that extra bit of variation in its pattern. Shapley applauded Sergei "for being so surprisingly lucky, or instinctively guided, that he can pull down such unusual eclipsing stars." Gaposchkin, by his own description, did not so much pull them down as "go fishing for stars" in the ocean of glass plates.

AT THE 1943 WINTER MEETING of the American Astronomical Society, held in Cincinnati, the executive council moved to award the fourth Annie Jump Cannon Prize to Miss Cannon's old friend and coworker Antonia Maury. The seventy-seven-year-old Miss Maury, a party to the discovery of spectroscopic binaries in 1889, had nurtured her interest in those stars through all the intervening years. She had also followed her pet variable, the strange Beta Lyrae, through hundreds of Harvard spectrograms taken over decades. She saw her account of "The Spectral Changes of Beta Lyrae" published in the Harvard *Annals* in 1933, two years before her retirement, but even now she continued visiting the observatory to review any new plates that included Beta Lyrae, whose behavior remained an enigma.

Miss Maury's aunt Antonia Draper Dixon died in 1923, leaving her niece the diamond ring that had once belonged to Anna Palmer Draper. The family property at Hastings-on-Hudson became a trusteeship of the American Scenic Historic Preservation Society. Miss Maury, who lived in the old observatory cottage built for her grandfather, wanted to turn part of the ten-acre tract into a botanical garden. She invited neighborhood children to wander freely about the grounds of "Draper Park," or accompany her on walks to learn the names of all the plants, birds, insects, and rocks that she had come to love in her own childhood. With guidance from Shapley, she purchased a secondhand Alvan Clark 6-inch telescope—not just for her personal use, but as an attraction for local residents, whom she would also treat to free public lectures in her areas of expertise. Members of the Hastings Amateur Astronomy Association laid a cement platform for the telescope in 1932, and a committee from the mayor's office raised funds for a shed to protect it. But the shed was never finished, and Miss Maury's grand design failed to materialize.

Recently Miss Maury had taken up the cause of the Western redwood forests. The wartime demand for lumber was feeding these trees to sawmills with no thought of their conservation, and she meant to alter the situation if possible. She spoke of it to Mrs. Gaposchkin, whom she regarded as the daughter she wished were hers, and who shared her appreciation of botany. Mostly they spoke of stars and spectra, but of enough other things besides to make Mrs. Gaposchkin describe Miss Maury as "a dreamer and a poet, always vehemently denouncing injustice, forever battling for a good (often a lost) cause."

Miss Maury's original stellar classification scheme received new recognition in 1943, when astronomers at the Yerkes Observatory proposed refinements to the Henry Draper Catalogue. The new "MKK" system, so named for William Morgan, Philip Keenan, and Edith Kellman, preserved Miss Cannon's letter categories in the accustomed order, and also the numbered subscripts, zero to nine, which graded intermediate spectral identities. The major innovation of the MKK system was the addition of Roman numerals I through V to designate each star's luminosity, or intrinsic brightness—the very quality that Miss Maury had attempted to characterize by her a, b, and c divisions. Morgan himself expressed the highest respect for Miss Maury, whom he considered an even more skilled stellar classifier than the late Miss Cannon.

The memorial *Annals* volume to be dedicated to Miss Cannon stalled during the war for lack of funds and personnel. By 1944 the number of observatory staff engaged in full-time war-related work had risen from twenty-five to thirty-two. Meanwhile good weather continued to smile on Bloemfontein, where the quantity of glass plates in storage tested the available space. Shapley yearned to see the fruits of the latest two years' labors in the Southern Hemisphere. Given the fairly reliable delivery of mail and supplies from England to Africa, and a sudden drop in insurance rates for international shipments, he asked Dr. Paraskevopoulos to send home some of the plates. About 1,500 photographs, or one-tenth of the accumulation, joined other cargo loaded aboard the *Robin Goodfellow,* bound from Cape Town to New York. On July 25, 1944, the ship was torpedoed in the South Atlantic, and sank, with the loss of all hands.

EVERYTHING LOOKED DIFFERENT after the war—after the atrocities typical of armed conflict had been overshadowed by the summary annihilation of several hundred thousand human beings by a new class of weapon. People began to speak of science as "having known sin."

Even as Shapley anticipated the return of the observatory's former good fortune, he recoiled from what he had seen. "Should we plan to construct large new buildings in urban areas in this age of atomic bombs?" he asked in his 1946 report to President Conant. "Should the staff of the Observatory, with its experience and specialized knowledge, help in the creation of international scientific institutions as a part of its contribution to international sanity? Should our experts in ballistics, rocket problems, optics, turn hopefully away from the wartime applications of science? Should we now work out a plan for burying our best photographs, records, and publications in such a way that they might be discovered and utilized in some later millennium when less social stupidity prevails among the higher animals?"

Such comments, paired with the director's liberal politics and aid to displaced foreign scientists, drew suspicion from the House Un-American Activities Committee. Shapley was subpoenaed by the committee in November 1946 to appear at a closed hearing in Washington, but he suffered no penalties as a result of that encounter. Later, when Senator Joseph McCarthy accused him of ties to Communist organizations, Shapley accused the senator of "telling six lies in four sentences, which is probably the indoor record for mendacity."

If the war had taught astronomers their fitness for national defense, it also taught the government the value of supporting certain areas of basic research in astronomy. The Sun, for example, was now known to affect the stratum of Earth's atmosphere where radio transmissions traveled. A high-altitude station to monitor the Sun's behavior that the Harvard Observatory set up in 1941 near Climax, Colorado, became the darling of the Office of Naval Research. During the war, when large-scale military operations depended on radio communication, attacks had been scheduled on the Sun's timetable. Advances in the new field of solar-terrestrial relations offered direct benefits to postwar commercial shipping and aviation. At the Climax site, a peacetime Harvard project to photograph meteors was yielding much-desired information on atmospheric temperature, density, and drag.

Government agencies saw no gain, however, in probing the variable stars, or the structure of the Milky Way and its place among the other galaxies. Shapley thus ran into difficulties resuscitating his own chosen areas of interest. He urgently needed to hire new computers, but the low salaries for those positions looked even lower after the war, when inflation drove up prices and new industries paid higher wages. Realizing the need for nonmilitary agencies to bolster basic research, Shapley helped establish the National Science Foundation in the United States, and also took part in the creation of UNESCO, the United Nations Educational, Scientific and Cultural Organization.

In 1946 the Harvard administration reacted to Shapley's leftist political activities by restructuring the observatory hierarchy. Shapley retained the title of director but ceded control to a new Observatory Council, which included Bart Bok, Donald Menzel, Cecilia Payne-Gaposchkin, and Fred Whipple, a meteor and comet expert on staff since 1931. Bok was promoted to full professor and associate director, with supervision of Oak Ridge. Menzel was named chairman of the astronomy department and associate director for solar work. Mrs. Gaposchkin retained her title of Phillips Astronomer.

Some of the former employees of the observatory were willing to come back to work after the war at their former low salaries. One of these, Ellen Dorrit Hoffleit, returned in 1948 for the love of astronomy, at half her army pay. Dr. Hoffleit, a 1928 graduate of Radcliffe, had moved into the observatory straight out of college. She started on variable stars, but soon switched to meteors and later to the determination of stellar brightness by the widths of spectral lines. Her wartime work had taken her from the MIT Radiation Laboratory to the Ballistic Research Laboratory of the army's Aberdeen Proving Ground in Maryland, to the White Sands Missile Range in New Mexico. She had computed everything from firing tables for navy cannons to the velocities of captured V-2 rockets. Now that she was once again sighting objects native to the sky, she availed herself of help from rented IBM computing equipment in analyzing data on stellar distribution. The days of the human computer were numbered—by zeros and ones.

FORMER PICKERING FELLOW Helen Sawyer Hogg took the news of her Annie Jump Cannon Prize not exactly in stride. Pleasure jostled with anxiety and the lethargy that had come over her in the past few months. "All Spring I have felt very doleful," she told Shapley in a letter dated July 25, 1949. She had seen him just recently at the June gathering of the American Astronomical Society, in Ontario, where she now lived. "I left the Ottawa meetings more depressed than when I went; and the night observing which I have been tackling systematically since my return has served only to convince me once more that I cannot fit in night work with my heavy family responsibilities. In other words, I seem to have reached the end of my tether." Helen and her husband, Frank, a native Canadian, had moved to Victoria, British Columbia, in 1931 to work at the Dominion Astrophysical Observatory. Only Frank had a job there, but Helen also worked full-time as a volunteer. She was the first woman allowed to use the 72-inch reflector. When the Hoggs' daughter, Sally, was born in 1932, Helen continued her observing runs with Sally in a basket beside her. Sympathetic observatory director John Stanley Plaskett got Mrs. Hogg a $200 grant, so she could hire a housekeeper to mind the baby. In 1935 Frank was offered a professorship at his alma mater, the University of Toronto, and the family moved back East. Helen, too, was hired in Toronto. She became a research officer in the astronomy department and the university-affiliated David Dunlap Observatory in 1936, the year of David Hogg's birth. The Hoggs had another boy, James, in 1937, and Helen published her "Catalogue of 1116 Variable Stars in Globular Clusters" in 1939. The outbreak of war gave her the opportunity, in 1941, to teach astronomy classes at the university, which she had continued ever since. "I have asked Frank to get me an indefinite leave of absence from my university position here, but he is very much upset at the thought." The Annie Jump Cannon Prize seemed to add a new weight of obligation. "In my opinion, this award carries with it a certain amount of responsibility, when made to a person my age, that is. [She was forty-four.] In other words, it does not look so good to take the award and quit!" Conflicted, she had not yet responded to the AAS secretary, C. M. Huffer, regarding her expected acceptance. "It has probably not crossed his mind that circumstances might make it advisable for me to refuse the award."

Shapley himself felt dispirited by his current distance from active research, but he was as capable as ever of heartening a former student, especially one who shared his long-term devotion to globular clusters. "There is little doubt but that you are undertaking too much in running a family at this critical stage," he replied on July 29, 1949, "and doing everything else. A leave of absence from the University work is obviously a good idea; but a study, with astronomical literature in it, and some photographs of clusters and the computing machine—that should not be given up, even if it must be established in one corner of some room at home. And also probably there is some interesting and not too laborious writing about old books* that should be done, just to keep the finger in the game until strength and time are less expensive. About that award—don't be silly, even if the weather is hot. The award is made for past accomplishments, and carries with it no responsibility for future activities. Suppose I should commence turning in medals because I have degenerated into being just a blank, blank director, personality smoother, instigator of labors by others. Let's both cheer up. One particular reason for such a resolve is that after fifteen or twenty lectures on cosmogony in the Harvard Summer School I have convinced myself that this is unquestionably the best universe I know of."

Shapley had introduced the graduate summer program in astronomy and astrophysics in 1935. It collapsed during the war, but roared back to life afterward, and currently enrolled more than a dozen students. Just as Pickering had made the observatory's name synonymous with photometry and photography, Shapley tied it securely to graduate education. He had fostered a generation of Harvard astronomers.

Mrs. Hogg accepted her Annie Jump Cannon Prize at the June 1950 meeting of the AAS, held at Indiana University in Bloomington. Not long afterward, on New Year's Day 1951, her husband, then the forty-six-year-old director of the David Dunlap Observatory, died of a heart attack. She assumed many of his professional duties, including the teaching of his courses and the writing of his weekly astronomy column for the *Toronto Star*, but not the directorship, which went to someone else.

In August 1951 Shapley gave notice that he would retire as director of the Harvard Observatory at the end of one more year, shortly before his sixty-seventh birthday. To his chagrin, the university hesitated to name a successor, either from within the ranks or from another institution. Months passed in which the staff grew increasingly insecure. At the same time, the lack of a designated new leader diminished the observatory's standing in

* "Old books" was her term for the historical astronomical catalogues and other texts that she discussed in her regular column, "Out of Old Books," for the *Journal of the Royal Astronomical Society of Canada*.

the eyes of prospective students and astronomers generally. In March 1952 President Conant appointed an ad hoc committee, chaired by his wartime colleague J. Robert Oppenheimer, to evaluate the entire Harvard astronomy program. As Shapley prepared to vacate his position that August, Donald Menzel of the Observatory Council was named temporary acting director.

Menzel shepherded the observatory through the ensuing period of upheaval. The next two years saw the demolition of the old wooden structures, the erection of brick office buildings alongside the Great Refractor, the eviction of the American Association of Variable Star Observers from the observatory grounds, and the abandonment of the Boyden Station in South Africa. Menzel was officially named the sixth director in January 1954, and in 1955 the Harvard College Observatory formed a beneficial new association with the Smithsonian Astrophysical Observatory, which moved from Washington, D.C., to Cambridge. At the Oak Ridge site, renamed the Agassiz Station in memory of patron George R. Agassiz, a large new telescope marked Harvard's entry into the emerging science of radio astronomy. Where a 60-inch optical reflector had once been dominant, an antenna 60 feet in diameter now collected faint radio signals from deep space.

Cecilia Payne-Gaposchkin became a full professor in 1956, the first woman at Harvard to be promoted to that level. She sent handwritten invitations to all the female astronomy students to join her for a celebration in the observatory library, where she pulled herself up to her full height, squared her broad shoulders, and said with a twinkle, "I find myself cast in the unlikely role of a thin wedge."

As full professor, she was eligible to become department chairman, which title was thrust on her the following fall. Although she had long craved the prestige of that position, the cares of office alternately bored her and strained her nerves. What was worse, they took time away from her research.

In 1958 the Harvard Corporation under President Nathan M. Pusey at last elected Cecilia Payne-Gaposchkin the Phillips Professor of Astronomy. Even then her salary of $14,000 a year, while higher than her husband's, remained far below that of her male peers.

• • •

CATHERINE WOLFE BRUCE's investment in astronomy came too late in life to answer her questions about the universe. The medal she endowed, however, continues today to attach her name to every significant advance in her adopted science. Among the more than one hundred Bruce gold medalists who have been celebrated for their lifetime achievements, Arthur Stanley Eddington deciphered the internal structure of the stars, realizing that their mass at the time of their birth determines their ultimate fate; Henry Norris Russell charted the course of stellar evolution, showing that stars do change from one color to another as they age; and Hans Bethe explicated the process of nuclear fusion by which stars generate their heat and light. In addition to Edward Pickering, Harvard College Observatory winners of the Bruce Medal include Harlow Shapley, Bart Bok, and Fred Whipple, who propounded the "dirty snowball" model of comet construction.

To date, only four women have received the Bruce Medal. The first, in 1982, was Margaret Peachey Burbidge, a native of England who studied the spectra of galaxies and, in collaboration with her husband, Geoffrey, and their colleagues William Fowler and Fred Hoyle, showed that all heavy elements are produced in the interiors of stars. In 1990 the Bruce Medal went to Charlotte Moore Sitterly. As Princeton computer Charlotte Moore in 1929, she took advantage of Henry Norris Russell's absence on sabbatical to enroll at the University of California, Berkeley, where she earned her Ph.D. in 1931, doing research on the spectra of sunspots. After she returned to Princeton and married astronomer Bancroft Sitterly in 1937, she continued working, and later became manager of the program in atomic spectroscopy at the National Bureau of Standards. Vera Rubin, who attended Vassar College because of its historic association with Maria Mitchell, received the 2003 Bruce Medal for her measurements of galaxy rotation, which led to the discovery of dark matter. Sandra Moore Faber, the 2012 winner, did her graduate work at Harvard but has spent her career at the University of California Observatories, pursuing the formation, structure, and clustering of galaxies. In 2013 she was one of twelve recipients of the National Medal of Science.

The telescope named for Miss Bruce, which Shapley praised as "the great galaxy hunter of the Southern Hemisphere," was decommissioned in 1950. It gave up its mount at Bloemfontein to a new 30-inch instrument that promised to provide even better photographs in shorter exposure times. The intact Bruce lens and tube stood idle for several years in Africa before being shipped back to the States, for continued idleness at Oak Ridge. The Bruce's old dome in Arequipa was converted to a chapel.

Miss Bruce herself lies buried, as she prearranged, at Green-Wood Cemetery in Brooklyn, New York, the final resting place of the city's wealthiest, most prominent citizens of her day. Dr. and Mrs. Henry Draper are also buried there, together, under a joint pentagonal marker with a carved facsimile of the congressional medal commending Dr. Draper's part in the 1874 transit of Venus.

The Draper Medal, like the Bruce Medal, continues to acknowledge astronomers for lifelong accomplishments. Those researchers who have won both the Draper and the Bruce include Edward Pickering, George Ellery Hale, Arthur Stanley Eddington, Harlow Shapley, and Hans Bethe. No women have ever taken both prizes. In the years since Miss Cannon was awarded the Draper Medal, only one other woman has received it—radio astronomer Martha P. Haynes of Cornell University, who shared the 1989 honor with Riccardo Giovanelli for their joint mapping of the large-scale distribution of galaxies.

The Annie Jump Cannon Prize likewise endures. It was awarded to Miss Cannon's former recorder, Margaret Walton Mayall, in 1958, and to Nantucket observatory director Margaret Harwood in 1961. The frequency of selection has increased since 2006, when the American Astronomical Society began choosing a new winner every year. The annual cash award now exceeds $1,000 (the amount originally contributed by Miss Cannon), but is no longer accompanied by a handmade pendant. In 2016 Laura A. Lopez of Ohio State University won the Cannon Prize for her studies in radio and X-ray astronomy regarding the life cycles of stars.

On Observatory Hill in Cambridge, Massachusetts, today, the Harvard-Smithsonian Center for Astrophysics stands as the successful union of the former Harvard and Smithsonian observatories. The CfA employs three hundred scientists engaged in university- and government-supported research covering every area of astronomy. Approximately one-third of the staff is female.

THE MONUMENTAL WORK of stellar classification known as the Henry Draper Catalogue and Extension, begun under Williamina Fleming in the 1880s and continued through 1940 by Annie Jump Cannon, is still in regular use. Every astronomy student learns the temperature order of the stars by memorizing Oh, Be A Fine Girl/Guy, Kiss Me. A contest to come up with a cleverer, less sexist mnemonic was held for several years in Harvard's introductory astronomy course, but the anonymous original retains its utility and pride of place. The thousands of Henry Draper identification numbers, assigned to the stars by the female computers, remain in effect as well. Star number HD 209458, for example, a variable in the constellation Pegasus, made news when modern detection methods located a planet in orbit around it.

Antonia Maury's classification system, with its twenty-two spectral types and several subtypes, struck her contemporaries as too complex to gain traction. Some of its distinctions proved crucial, however, in discerning the different magnitudes and ages of stars that shared the same general categories. After Ejnar Hertzsprung first complimented Miss Maury's acumen in 1908, the Draper classification made room for one of her notations in 1922, and in 1943 the MKK innovation incorporated additional Maury-type gradations. In 1978, some twenty-five years after her death, her system won further vindication when William Morgan published the *Revised MK Spectral Atlas for Stars Earlier Than the Sun* with new coauthors Helmut Abt and J. W. Tapscott. Morgan dedicated this volume "To Antonia C. Maury (1866–1952) Master Morphologist of Stellar Spectra."

Henrietta Leavitt did not participate in the classification effort, but her pursuit of variable stars and her discovery of the relationship between period and brightness among the Cepheid variables has had an equal, if not greater, impact on progress in astronomy. Once calibrated and applied to the problem of measuring distances across space, Miss Leavitt's period-luminosity relation allowed Harlow Shapley to extend the boundaries of the Milky Way. The same Cepheid stars, subjected to the same analytical techniques, enabled Edwin Hubble to appreciate the enormous distances to the spiral nebulae. Hubble used Cepheids in 1924 to show that the Milky Way was not the only galaxy in the universe, and later to demonstrate that the universe was expanding to ever vaster proportions, as evidenced by the speedy outbound flight of most external galaxies. The Cepheids, however, had still more to say about cosmic distances. During World War II, Walter Baade, a German immigrant who had been working at Mount Wilson since 1931, took advantage of dark skies made darker by area-wide blackouts. Baade's detailed study of the stars of the Andromeda Galaxy split the Cepheids into two subgroups. He accordingly recalibrated the distance scale and arrived at an overall size of the universe that doubled Hubble's estimate. Today, astronomers rely on the period-luminosity relation to measure the current expansion rate of the universe.

The relationship between redshift and distance that Hubble saw in the realm of the nebulae has come to be known as Hubble's law. By the same token, some scientists argue, the relationship between period and brightness that provided the basis for Hubble's discoveries should rightly be renamed the Leavitt law. Awareness of this proposed terminology has been spreading since January 2009, when the executive council of the American Astronomical Society unanimously passed a resolution in favor of the change. The occasion was the one-hundredth anniversary "of Henrietta Leavitt's first presentation of the Cepheid Period-Luminosity relation, a seminal discovery in astronomy that continues to have great significance." Although the councilors allowed that the AAS had "no authority to define astronomical nomenclature," they said that they personally "would be very pleased" to see the designation "Leavitt Law" in wide use.

When the female computers of the Harvard College Observatory come up in present-day conversation, they are often portrayed as underpaid, undervalued victims of a factory system. Pickering stands accused of giving them scut work that no man would stoop to do, yet this is far from true. Before astronomy morphed into astrophysics around the turn of the twentieth century, both men and the few women engaged in the science were willing slaves to routine. Arthur Searle, the acting director during the interregnum between Winlock and Pickering, tried to explain this reality to a journalist intent on chronicling the excitement of observatory life. "It is only fair to warn you," Searle admonished Thomas Kirwan of the *Boston Herald*, "that your proposed article cannot be at once true and entertaining. The work of an astronomer is as dull as that of a book-keeper, which it closely resembles. Even the results reached by astronomical work, although they relate to more dignified subjects than the ordinary affairs of trade, are far less interesting than the result of book-keeping, at least to the general reader, unless they are so disguised by fancy as to have little to do with science."

Pickering, though enthralled by the incremental gains he could make nightly at the controls of his photometer, ushered in a new era of photography and spectroscopy that transformed the observatory. Having found several female assistants already in place when he took charge, he brought in more of them and entrusted the stellar classification to their judgment. He also attracted assistance in variable star observation from alumnae and female professors of the women's colleges. His treatment of women, widely perceived as more than fair, invited fellowship funding that further advanced women's participation in astronomy. When Harlow Shapley came to Harvard, he was able to redirect the fellowship money into a program of graduate education that initially—and necessarily—favored women over men as applicants. Cecilia Payne's attainment of the first astronomy Ph.D. at Harvard, in the course of which she challenged the very fabric of the universe, could be traced directly to Pickering's "harem" and the observatory's singular collection of glass plates.

NO ASTRONOMER WORKING TODAY uses glass plates to photograph the cosmos. CCDs, or charge-coupled devices, began replacing photographic film in the 1970s, and for the past two decades virtually all celestial images have been captured and stored digitally. But no matter how broadly or deeply modern sky surveys probe outer space, they cannot see what the heavens looked like on any given date between 1885 and 1992. The record preserved in the Harvard plate collection of one hundred years of starry nights remains unique, invaluable, and irreplaceable.

The half a million glass plates reside in the expanded Brick Building. They stand on their long edges, leaning slightly to the left or right on the shelves of the many metal cabinets. Some early photographs still wear their original paper jackets, covered over with handwritten commentary from their long-ago keepers. Old or new, each envelope bears an affixed bar-coded sticker containing accession information that helps the current curator maintain order in the plate stacks. Visiting researchers file in and out. Historians prize the plates for their dated information, for the antiquated union of glass and silver-gelatin emulsion that embeds the stars. Astrophysicists consult the plates to enrich and interpret the latest findings through "time domain astronomy." Celestial denizens undreamed of at the start of Pickering's sky patrol—pulsars, quasars, black holes, supernovae, X-ray binaries—nevertheless left their marks on the plates.

When computers were human, they scanned these photographs by eye for as many interesting objects as they could find. There were never enough "readers" to utilize the plate library to Pickering's or Shapley's satisfaction. The most motivated of their methodical workers, when confronted with an image containing as many as one hundred thousand stars, could carry discovery only so far. Even now, the information content of the plate stacks is largely untapped.

To extract all that waiting data with modern computerized algorithms, the Harvard-Smithsonian Center for Astrophysics inaugurated a digitization project in 2005, with funding from the National Science Foundation. The ongoing goal is to clean, scan, and analyze every plate, so as to provide "Digital Access to a Sky Century @ Harvard," or DASCH. After more than ten years, the work is approximately one-quarter complete.

All procedures and instruments for DASCH (pronounced "dash") have had to be invented and assembled on-site, from the ingenious machine that cleans the plates with ethanol solution to the high-speed scanner that accommodates the standard 8 × 10 and also the Bruce-size 14 × 17 plates. At each stage of activity, curatorial concerns vie with scientific requirements. For example, the plate-cleaning process, an essential prelude to producing clear, crisp scans, automatically erases any markings jotted on the glass by iconic figures such as Henrietta Leavitt and Annie Cannon. The compromise solution is to photograph each marked photograph before cleaning it—and each jacket, too—to record all those notations. Certain plates are judged too historic to be tampered with, and will be archived indefinitely. One of these holds an image of a star field made while the nature of spiral nebulae still sparked debate. On it, someone circled a tiny swirl of matter too small to see without a magnifying loupe. Next to the inked circle, an inked question arises: *Galaxy?*

The index cards and logbooks that list the telescope, sky coordinates, date, and exposure time of each photograph are also coming online, thanks to willing individuals who spend a few hours every day transcribing them via the Smithsonian Institution's crowdsourcing platform. Citizen scientists work in front of their own computer screens, from high-resolution photographs of each logbook's hundred pages, each page crammed with statistics and remarks on as many as twenty plates.

At the outset, DASCH team members named reasons beyond data mining as justification for their long-term project. They wanted to make the plates available for convenient worldwide use, to protect them from careless handling by interested borrowers, and to save the contents from predictable deterioration, such as emulsion separation. Once the process was under way, an unanticipated event provided further justification for the effort.

On Monday morning, January 18, 2016, a water main burst under a courtyard at 60 Garden Street, the official address of the CfA. The pipe provided water to four nearby buildings, including the original Brick Building and its 1902 and 1931 extensions. The rupture released water underground with enough force to breach the foundation walls and flood the lower level of the plate vault. Approximately sixty-one thousand plates were submerged. Experts from the on-campus Weissman Preservation Center responded to the emergency and diagnosed mold as the worst-case outcome of immersion. Spores that colonized the plates might configure their own new biological constellations. For all of Pickering's foresight in initiating and protecting the collection, he never suspected that water, not fire, would threaten its integrity.

The conservators prescribed the immediate removal of the plates to a dry place where they could be kept well under zero degrees Centigrade—too cold for mold to grow. The prevailing weather conditions at the time, clear with temperatures below freezing, turned the outdoors into a temporary safe haven. As soon as all the water had been pumped out of the building on Monday, dozens of volunteers came to the collection's aid; they traipsed in and out of the stacks all through Tuesday night and Wednesday, carrying armloads of fragile plates to dry ground. Not a single piece of glass was broken.

By Thursday the rescued plates had been driven in trucks to the Polygon Document Restoration Services in North Andover, where they were put into frozen storage, to be later thawed out and cleaned, one by one.

One by one, the way the stars emerge as evening falls, the drowned, muddied plates will revive the vivid skyscapes that impressed them when they were sensitive to light. Once again they will reveal the stellar spectra, the variable stars, the star clusters, the spiral galaxies, and all the other luminous sights they first conveyed to a small but dedicated circle of women.

APPRECIATION

My warmest thanks to:

Wendy Freedman, the John & Marion Sullivan University Professor of Astronomy and Astrophysics at the University of Chicago, who planted the idea for this book more than twenty years ago;

Michael Carlisle of InkWell Management, who helped shape the project and find it an ideal home with Kathryn Court at Viking;

Alison Doane, Jonathan Grindlay, David Sliski, and Lindsay Smith, for access to the glass universe in the Harvard Plate Stacks;

Christopher Erdmann, Maria McEachern, Amy Cohen, Louise Rubin, Katie Frey, and Daina Bouquin of the John G. Wolbach Library in the Harvard-Smithsonian Center for Astrophysics, for taking me in as one of their own;

Robin McElheny, Tim Driscoll, Pamela Hopkins, Robin Carlaw, Barbara Meloni, Ed Copenhagen, Caroline Tanski, Samuel Bauer, Michelle Gachette, and Jennifer Pelose in the Harvard University Archives, for opening Miss Cannon's diaries and other paper treasures;

Susan Ware, Sarah Hutcheon, and Jane Kamensky of the Schlesinger Library, for illuminating the Radcliffe backgrounds of numerous Harvard ladies;

Smith College students, faculty, and staff, for providing the perfect environment in which to write about women's history;

William Ashworth, Barbara Becker, David DeVorkin, Suzan Edwards, Owen Gingerich, Alyssa Goodman, Katherine Haramundanis, Doug Offenhartz, Jay and Naomi Pasachoff, William Sheehan, Joseph Tenn, and Barbara Welther, for reading and commenting on early drafts;

Thomas Fine and Lia Halloran, for help illustrating this story;

Isaac Klein, Stephen Sobel, Alfonso Triggiani, Barry Gruber, and Gary Reiswig, for their consistent encouragement;

Sheryl Heller and the crew at GeekHampton in Sag Harbor, New York, for invaluable assistance with the *other* kind of computers.

SOURCES

CHAPTER ONE: Mrs. Draper's Intent

The letters between Anna Palmer Draper and Edward Pickering are held in the Harvard University Archives, along with all the other observatory correspondence, and are quoted here with permission.

Pickering's call for women assistants in the observation of variable stars was announced in the *Statement of Work Done at the Harvard College Observatory During the Years 1877–1882* and was also issued as a separate brochure, "A Plan for Securing Observations of the Variable Stars."

Hard copies of all the observatory publications, such as the *Annals* and the annual reports, are held in the John G. Wolbach Library at the Harvard-Smithsonian Center for Astrophysics in Cambridge. Most of these materials have been digitized and can be read online at http://adsabs.harvard.edu/historical.html.

CHAPTER TWO: What Miss Maury Saw

The letters of Antonia Maury to her Draper relatives are held in the Library of Congress and excerpted here with the permission of her family.

All correspondence concerning the Boyden Station of the Harvard College Observatory is held in the Harvard University Archives.

CHAPTER THREE: Miss Bruce's Largesse

Letters to Edward Pickering from Catherine Wolfe Bruce, as well as those from her sister Matilda, are held in the Harvard University Archives.

The article by astronomer Simon Newcomb that galvanized Miss Bruce was titled "The Place of Astronomy Among the Sciences"; it appeared in 1888 in the *Sidereal Messenger.*

Williamina Fleming's prepared speech for the Chicago conference was published as "A Field for Woman's Work in Astronomy" in 1893 in *Astronomy and Astro-Physics.*

CHAPTER FOUR: Stella Nova

Edward Pickering reported Mrs. Fleming's first nova discovery, "A New Star in Norma," in the pages of *Astronomy and Astro-Physics.*

Pickering's correspondence with Antonia Maury, also with the Reverend Mytton Maury, is held in the Harvard University Archives.

CHAPTER FIVE: Bailey's Pictures from Peru

Annie Jump Cannon was a lifelong diarist and prolific letter writer. Her diaries, scrapbooks, and other papers, including the libretti she collected for the many opera performances she attended, are held in the Harvard University Archives.

Antonia Maury's "Verses to the Vassar Dome," written in 1896, were printed in *Popular Astronomy* in 1923.

Edmond Halley summoned astronomers to observe the transit of Venus with his announcement, in Latin, of "A New Method of Determining the Parallax of the Sun," published in *Philosophical Transactions of the Royal Society* in 1716.

CHAPTER SIX: Mrs. Fleming's Title

The handwritten Journal of Williamina Paton Fleming, part of the Harvard "Chest of 1900," is held in the University Archives and can be read online at http://pds.lib.harvard.edu/pds/view/3007384.

President Edward B. Knobel's comments pertaining to the presentation of Edward Pickering's second gold medal were published in *Monthly Notices of the Royal Astronomical Society* in February 1901.

CHAPTER SEVEN: Pickering's "Harem"

The correspondence between Andrew Carnegie and Edward Pickering, also the letters exchanged by Louise Carnegie and Williamina Fleming, are held in the Harvard University Archives.

CHAPTER EIGHT: Lingua Franca

Herbert Hall Turner commented on the "marvellous" achievements of Williamina Fleming in the obituary he wrote for her, which was published in *Monthly Notices of the Royal Astronomical Society* in 1911.

Edward Pickering's diary of his trip to Pasadena to attend the 1910 meeting of the Solar Union is held in the Harvard University Archives, and was transcribed for publication in *Southern California Quarterly* by historian Howard Plotkin.

CHAPTER NINE: Miss Leavitt's Relationship

Frank Schlesinger of the Allegheny Observatory collated the responses of all the astronomers polled after the 1910 meeting in Pasadena and published their comments in the *Astrophysical Journal* under the heading "Correspondence Concerning the Classification of Stellar Spectra."

CHAPTER TEN: The Pickering Fellows

Harlow Shapley reminisced about his life experiences in a breezy memoir called *Through Rugged Ways to the Stars,* published in 1969. He dedicated the book "To the memory of Henry Norris Russell."

Margaret Harwood's letters to Annie Jump Cannon, Edward Pickering, and Harlow Shapley are preserved in the Harvard University Archives along with other materials pertaining to the observatory, but most of her private papers and photographs are held at the Schlesinger Library on the History of Women in America, at the Radcliffe Institute for Advanced Study, Cambridge.

CHAPTER ELEVEN: Shapley's "Kilo-Girl" Hours

As chairman of the Astronomical Fellowship Committee, Annie Jump Cannon chronicled the activities of current and past Pickering Fellows in her write-ups for the Nantucket Maria Mitchell Association Annual Report. These can be read online, courtesy of the Smithsonian Astrophysical Observatory/NASA Astrophysics Data System, at http://www.adsabs.harvard.edu.

CHAPTER TWELVE: Miss Payne's Thesis

Cecilia Payne recounted her life experiences in a treatise she named "The Dyer's Hand." It was published in 1984, along with admiring essays by several of her colleagues, in *Cecilia Payne-Gaposchkin: An Autobiography and Other Recollections,* edited by her daughter, Katherine Haramundanis.

CHAPTER THIRTEEN: *The Observatory Pinafore*

Helen Sawyer Hogg recalled the events that shaped her astronomy career when she spoke at a commemorative symposium held August 25–29, 1986, in Cambridge. The proceedings were later published as a book, *The Harlow Shapley Symposium on Globular Cluster Systems in Galaxies,* edited by Jonathan E. Grindlay and A. G. Davis Philip.

The proceedings of another symposium, organized in 2000 to mark the one hundredth anniversary of Cecilia Payne's birth, were published in *The Starry Universe: The Cecilia Payne-Gaposchkin Centenary,* edited by A. G. Davis Philip and Rebecca A. Koopmann. This volume includes lyrics sung by "Josephine" and other characters from *The Observatory Pinafore.*

CHAPTER FOURTEEN: Miss Cannon's Prize

The comments of Radcliffe dean Bernice Brown and Astronomer Royal Sir Frank Dyson at the opening of the 1932 IAU General Assembly were published in *Transactions of the International Astronomical Union,* volume 4.

CHAPTER FIFTEEN: The Lifetimes of Stars

The American Astronomical Society executive council's resolution regarding recognition of Henrietta Leavitt, including the officers' wish to see the Cepheid period-luminosity relation renamed the Leavitt law, was published in the *AAS Newsletter* for May/June 2009. The new term originated in 2008, during a conference held at the Harvard-Smithsonian Center for Astrophysics to celebrate the centenary of Miss Leavitt's discovery. Information about the conference, with links to all the presentations, can be found at https://www.cfa.harvard.edu/events/2008/leavitt.

SOME HIGHLIGHTS IN THE HISTORY OF THE HARVARD COLLEGE OBSERVATORY

1839 Observatory established by Harvard Corporation at Dana House.

William Cranch Bond appointed astronomical observer.

1843 Great Comet's visit inspires citizens of Boston and environs to fund the purchase of a large telescope for the observatory.

1844 Observatory moves to Summerhouse Hill, where a suitable foundation is laid for the new 15-inch telescope.

1845 First Harvard College Observatory Visiting Committee is headed by John Quincy Adams.

1846 George Phillips Bond appointed assistant observer.

First annual report issued.

1847 The "Great Refractor," a 15-inch telescope with a lens crafted in Munich, is mounted in the new observatory building.

1848 The Bonds discover the eighth moon of Saturn and name it Hyperion.

Edward Bromfield Phillips bequeaths $100,000 to the observatory, to pay salaries and all operating expenses.

1849 Statutes make the observatory a department of the university, and change the elder Bond's title to director.

1850 First photograph of a star, Vega, taken by George Phillips Bond and John Adams Whipple.

Jenny Lind sees a meteor fireball through the Great Refractor.

1856 First volume of the *Annals of the Astronomical Observatory at Harvard College* published.

1859 At death of William Cranch Bond, George Phillips Bond becomes second director.

1866 Joseph Winlock appointed as third director.

1868 Arthur Searle joins staff as assistant.

1870 Meridian circle, an instrument for ascertaining star positions, is built for Winlock in London and installed at Harvard.

William Rogers takes charge of meridian observations for astrometry (star positions).

1875 At Joseph Winlock's death, his daughter Anna joins computing staff.

Miss Rhoda G. Saunders is hired as the first female computer from outside the observatory family.

1876 Arthur Searle serves as interim director.

1877 Edward Charles Pickering takes office as fourth director, initiates his program of stellar photometry.

1879 Williamina Fleming is hired as a maid in the Pickering household.

Edward Pickering introduces the meridian photometer for judging the brightness of stars.

1880 Edward Pickering publishes his five-type classification of variable stars.

1881 Williamina Fleming becomes a permanent member of the observatory staff.

1882 Edward Pickering and his brother William, of MIT, experiment with lenses for photographing the night sky.

Director Pickering issues a call for volunteers, especially women, to observe variable stars and share their results with Harvard.

1883 Harvard Observatory becomes the designated dispenser of information regarding cometary and other discoveries, made by observers anywhere, and telegraphed to observatories everywhere.

1884 Results of first photometry study published in the *Annals*, vol. 14.

Edward Pickering divides entire sky into forty-eight equal regions known as the Harvard Standard Regions.

1885 Bache Fund grant provides the 8-inch telescope required for Pickering's program of nightly sky photography.

Williamina Fleming begins to measure and compute stellar magnitudes from photographs.

1886 Anna Palmer Draper provides funding for photography of stellar spectra, with the aim of realizing the unfulfilled dream of her late husband, Dr. Henry Draper.

Edward Pickering receives the gold medal of the Royal Astronomical Society in recognition of the Harvard Photometry.

1887 Harvard acquires the Boyden Fund to build a high-altitude observatory.

William Pickering joins the observatory staff.

Edward Pickering is named the Paine Professor of Practical Astronomy; Arthur Searle assumes the title of Phillips Professor.

1888 Antonia Maury joins the staff of female computers, begins to study the spectra of bright northern stars.

1889 Solon Bailey begins observations in Peru, aided by his wife, Ruth E. Poulter Bailey.

Catherine Wolfe Bruce gives $50,000 for construction of a 24-inch astrophotographic telescope.

Edward Pickering discovers the first spectroscopic binary; Antonia Maury finds the second one.

1890 "The Draper Catalogue of Stellar Spectra" is published in the *Annals*, vol. 27, with classifications made by Williamina Fleming.

Solon Bailey establishes Harvard's Boyden Station at Arequipa.

1891 William Pickering takes over in Arequipa as Boyden Station director.

Arthur Searle begins teaching astronomy classes for women.

1893 Solon Bailey resumes charge of Boyden Station in Peru.

Glass plates are moved to new fireproof Brick Building.

Williamina Fleming prepares "A Field for Woman's Work in Astronomy" for presentation at the Columbian Exposition in Chicago; discovers her first nova on plates from Arequipa.

Bruce telescope sees first light at Cambridge.

1895 Edward Pickering institutes the *Harvard College Observatory Circular* to describe news of the observatory, beginning with Williamina Fleming's discovery of Nova Carinae (her second nova) from photographs taken at Arequipa; her third such discovery, Nova Centaurus, follows a few months later.

Henrietta Swan Leavitt volunteers at the observatory.

Solon Bailey discovers many variables within certain star clusters of the Southern Hemisphere.

1896 Annie Jump Cannon joins the observatory as a research assistant, commences her study of the spectra of bright southern stars.

Bruce telescope arrives at Arequipa.

1897 Antonia Maury publishes "The Spectra of Bright Stars" in the *Annals*, vol. 28, and is acknowledged as the author on the title page.

1898 National professional organization of astronomers, later named the Astronomical and Astrophysical Society of America, established at a meeting held at Harvard.

Edward Pickering introduces *Harvard College Observatory Bulletins* to augment the telegraphic announcements with details sent by mail.

1899 Williamina Fleming given Harvard title as curator of astronomical photographs.

William Pickering discovers a ninth satellite of Saturn, Phoebe.

1900 Harvard "Chest of 1900" time-capsule project invites Edward Pickering and Williamina Fleming to chronicle their daily activities.

Catherine Wolfe Bruce dies.

1901 Edward Pickering receives his second gold medal from the Royal Astronomical Society, for variable star studies and advances in astrophotography.

Annie Cannon publishes catalogue of bright southern stars in the *Annals,* vol. 28.

1903 Annie Cannon publishes her "Provisional Catalogue of Variable Stars" in the *Annals,* vol. 48.

After an absence of several years, Henrietta Leavitt returns as a full-time employee.

Edward Pickering issues "Photographic Map of the Entire Sky."

1905 Henrietta Leavitt notices an inordinate number of variables in the Magellanic Clouds.

Edward Pickering elected president of the Astronomical and Astrophysical Society of America.

1906 Edward Pickering and Henrietta Leavitt embark on a large-scale determination of photographic magnitudes.

Williamina Fleming elected to honorary membership in the Royal Astronomical Society.

1907 Annie Cannon publishes her "Second Catalogue of Variable Stars" in the *Annals,* vol. 55.

Williamina Fleming publishes "A Photographic Study of Variable Stars" in the *Annals,* vol. 47.

Margaret Harwood joins the staff.

1908 Edward Pickering publishes the Revised Harvard Photometry in the *Annals,* vols. 50 and 54.

Solon Bailey compiles a whole-sky catalogue of 263 bright clusters and nebulae in the *Annals,* vol. 60.

Henrietta Leavitt publishes her discovery of "1777 Variables in the Magellanic Clouds" in the *Annals,* vol. 60.

Edward Pickering receives the Catherine Wolfe Bruce Gold Medal.

1909 Solon Bailey reconnoiters potential new observatory sites in South Africa.

1910 Foreign astronomers attend meeting of the Astronomical and Astrophysical Society of America, held at Harvard.

International Union for Cooperation in Solar Research, meeting in Pasadena, adopts Harvard's Draper classification system developed by Annie Cannon.

1911 Williamina Fleming dies.

American Association of Variable Star Observers is founded by William Tyler Olcott, one of Pickering's volunteer contributors.

1912 *Harvard Bulletin* switches from handwritten and mimeographed production to printed format.

Edward Pickering and Annie Cannon demonstrate the brightness of B stars.

Henrietta Leavitt publishes her "period-luminosity relation."

Margaret Harwood becomes first Astronomical Fellow of the Nantucket Maria Mitchell Association.

Astronomical and Astrophysical Society of America renamed the American Astronomical Society (AAS).

Annie Cannon elected treasurer of AAS, its first female officer.

1913 Henry Norris Russell and Ejnar Hertzsprung independently arrive at the significant relation between absolute magnitude and spectral type (later named the Hertzsprung-Russell diagram).

1914 Annie Cannon becomes honorary member of the Royal Astronomical Society.

Margaret Harwood investigates the light curve of the asteroid Eros.

Anna Palmer Draper dies.

1915 Margaret Harwood is named director of the Maria Mitchell Observatory on Nantucket.

1916 Nantucket Maria Mitchell Association establishes the Edward C. Pickering Astronomical Fellowship for Women.

Solon Bailey completes a provisional catalogue of 76 globular clusters in the *Annals*, vol. 76.

1918 First of nine volumes of the greatly expanded Henry Draper Catalogue is published in the *Annals*, beginning with vol. 91.

1919 Edward Pickering dies.

Solon Bailey serves as interim director.

1920 Harlow Shapley and Heber Curtis debate the scale of the universe.

1921 Harlow Shapley is named fifth director.

Henrietta Leavitt dies.

Harlow Shapley and Annie Cannon explore the relation between spectral type and magnitude.

1922 International Astronomical Union adopts Harvard's Draper stellar classification, representing the work of Williamina Fleming, Antonia Maury, and especially Annie Jump Cannon.

1923 Adelaide Ames enrolls as Harvard's first graduate student in astronomy.

Cecilia Payne arrives from England as Harvard's second graduate student in astronomy.

Harvard Reprints series initiated to disseminate staff members' published articles in professional journals.

1924 Harlow Shapley issues the first in a series of papers detailing the distance, size, and structure of the Magellanic Clouds.

Ninth volume of the Henry Draper Memorial is published in the *Annals,* vol. 99.

1925 Harlow Shapley introduces a new publication series of books, the Harvard Monographs, beginning with Cecilia Payne's doctoral dissertation, *Stellar Atmospheres.*

1926 *Harvard Bulletin* switches to monthly publication, each issue containing several items of interest.

Harlow Shapley introduces Harvard Announcement Cards for news (of comets, novae, asteroids) between issues of the *Bulletin.*

1927 Number of known variable stars reaches five thousand, more than four thousand of which are Harvard discoveries, found on the glass plates.

Harlow Shapley and Helen Sawyer complete new catalogue of globular clusters, their number increased to 95.

Boyden Station moves from South America to South Africa.

1929 Priscilla Fairfield marries Bart Bok.

1930 Helen Sawyer and Frank Hogg are married.

1931 Solon Bailey dies in June, Edward King in September.

Annie Cannon receives the Draper Medal of the National Academy of Sciences.

1932 Adelaide Ames dies.

International Astronomical Union meets at Harvard.

1933 Antonia Maury publishes "The Spectral Changes of Beta Lyrae" in the *Annals,* vol. 84.

Several Harvard telescopes move to rural site at Oak Ridge.

1934 Cecilia Payne and Sergei Gaposchkin elope.

Cecilia Payne-Gaposchkin wins the Annie Jump Cannon Prize.

1935 Harlow Shapley inaugurates graduate summer program in astronomy and astrophysics.

1939 Annie Cannon finds Harvard's ten thousandth variable star.

1941 Annie Cannon dies.

1943 Antonia Maury receives the Annie Jump Cannon Prize.

1946 An Observatory Council, including Bart Bok, Donald Menzel, and Cecilia Payne-Gaposchkin, is appointed to advise the director on policies and programs.

1949 Margaret Walton Mayall completes the Henry Draper Extension, published as the Annie J. Cannon Memorial Volume in the *Annals,* vol. 112.

1950 Helen Sawyer Hogg wins the Annie Jump Cannon Prize.

1952 Antonia Maury dies.

Harlow Shapley retires.

Donald Menzel becomes acting director.

1954 Donald Menzel officially named sixth director of the observatory.

1955 Smithsonian Astrophysical Observatory moves from Washington, D.C., to Cambridge to collaborate with the Harvard College Observatory.

1956 Cecilia Payne-Gaposchkin becomes the first woman at Harvard promoted to full professor; she is also named chair of the astronomy department.

1973 Formation of the Harvard-Smithsonian Center for Astrophysics unites the two observatories under a single director.

1979 Cecilia Payne-Gaposchkin dies.

2005 Inauguration of plate digitization process, Digital Access to a Sky Century at Harvard (DASCH).

GLOSSARY

American Astronomical Society The first national professional society of astronomers in the United States, founded in 1898 and originally called the Astronomical and Astrophysical Society of America.

Astronomical unit The distance from Earth to the Sun.

Astronomische Gesellschaft The second oldest astronomical society (after London's Royal Astronomical Society), established in Heidelberg in 1863.

Binary star A pair of stars moving around a common center of gravity.

Brightness—see **magnitude**.

Cepheid A pulsating variable star that changes brightness in a characteristic, predictable cycle, making it useful in estimating cosmic distances.

Chromatic aberration Blur or haze, caused by the several colors of light coming to focus at different distances from a lens.

Circumpolar star A star that neither rises nor sets, but circles the celestial pole.

Clock drive A mechanical or electric device that moves the telescope counter to Earth's turning, allowing it to stay focused on a given object.

Cluster A group of associated stars.

Cosmogony A theory about the origin and evolution of the universe.

Declination The latitude measure of the heavens; that is, the angular distance of an object above or below the celestial equator (the projection of Earth's equator onto the sky).

Doublet A pair of lenses combined for a desired effect.

Eclipsing binary, or eclipsing variable A pair of associated stars orbiting a common center of gravity, oriented so they pass in front of each other in the observer's line of sight.

Electromagnetic spectrum The full range of stellar radiation, from the longest wavelength (radio waves) to the shortest (gamma rays).

Ephemeris A table of predicted positions for a celestial body such as a planet, a moon, or a comet.

Epoch A reference date chosen for astronomical observations.

Flash spectrum The sudden change of the solar spectrum lines from dark to bright in the moments just before (and immediately following) the total phase of a solar eclipse.

Fraunhofer line A dark absorption line in the continuous (rainbow-colored) spectrum.

Galaxy A system consisting of billions of stars plus abundant dust and gas.

Globular cluster A group of many thousand associated stars with a dense central concentration.

Ion An atom that has lost one or more electrons and has a positive charge.

Island universe A term originally coined by Immanuel Kant (1724–1804) to denote a star system similar to, but separate from, our own galaxy.

K line One of the dark absorption lines seen in the solar spectrum and many other stellar spectra; it indicates the presence of ionized calcium.

Light curve The graphic representation of a variable star's (or other celestial body's) changing brightness over time.

Luminosity The intrinsic brightness of a star, or the total amount of energy it emits per unit of time.

Magellanic Clouds Two dense conglomerations of stars and nebulae seen from the Southern Hemisphere, now known to be satellite galaxies of the Milky Way.

Magnitude The brightness of an object, as judged by various standards over the centuries. The higher the number, the dimmer the appearance. Astronomers distinguish between "apparent" magnitude, or the way the object appears to earthly observers depending on its distance, and "absolute" magnitude, its intrinsic brightness.

Messier numbers (M-31 and others) Identification labels introduced by comet hunter Charles Messier (1730–1817), who needed a way to keep track of nebulous objects that were *not* comets.

Metals The term astronomers apply to all elements heavier than hydrogen and helium.

Meteor A particle, often a bit of comet dust no bigger than a grain of sand, that enters Earth's atmosphere and burns up by friction, looking like a "shooting star."

Milky Way The bright band of starlight stretching across the sky that has meant many things to stargazers over the ages, from the spilled milk of the goddess Hera to the name of the home galaxy where our solar system resides.

Nebula At the start of this story, any blurred object in space; today, an enormous interstellar cloud of ionized gases.

North Polar Sequence The forty-six stars (later increased to ninety-six) chosen as standards of comparison for precise determinations of photographic magnitudes.

Objective lens The light-gathering lens of a telescope, at the opposite end from the eyepiece.

Open cluster A group of a few hundred associated stars.

Orion Nebula The bright object in the sword of Orion, the Hunter, designated M-42.

Parallax The shift, or difference in apparent position, of an object against its background when viewed from two separate vantage points. Astronomers use parallax measures to estimate distances up to a few hundred light-years from the Sun.

Period The time span in which a variable star cycles through its brightness changes.

Personal equation An astronomer's reaction time.

Proper motion Movement of a celestial body across the line of sight.

Radial velocity An object's speed of approach or recession along the line of sight.

Radio astronomy A complement to optical astronomy; the study of electromagnetic radiation at wavelengths much longer than those of visible light.

Redshift The shift of known spectral lines toward the red end of the spectrum, caused by the object's motion away from the observer.

Right ascension The celestial equivalent of longitude for stating star positions.

Royal Astronomical Society The world's first organization of astronomers, founded in 1820 as the Astronomical Society of London.

Seeing The quality of the observing conditions, ideally cloudless skies and minimal movement of air. Astronomers rate seeing on a scale ranging from one (very poor) to ten (perfect).

Spectrum The rainbow of colors (and Fraunhofer lines) contained in visible light.

Spiral nebula An early term for a spiral galaxy.

Visible light A small portion of the electromagnetic spectrum, bordered by infrared and ultraviolet rays.

A CATALOGUE OF HARVARD ASTRONOMERS, ASSISTANTS, AND ASSOCIATES

George Russell Agassiz (July 21, 1862–February 5, 1951), like his famous father and grandfather, held a faculty position with Harvard's Museum of Comparative Zoology. He became an influential and generous member of the observatory's Visiting Committee. After his death, his wife, Mabel Simpkins Agassiz, continued that generosity.

Adelaide Ames (June 3, 1900–June 26, 1932), a Vassar alumna, was the observatory's first graduate student in astronomy, earning her master's degree from Radcliffe in 1924. She worked with Director Harlow Shapley to catalogue galaxies.

Solon Irving Bailey (December 29, 1854–June 5, 1931) extended the reach of the observatory by reconnoitering good locations for high-altitude satellite stations, first in South America and later in South Africa. He identified and studied the variable stars in globular clusters, which he called "cluster variables."

Bartholomeus Jan Bok (April 28, 1906–August 5, 1983) chose the structure and evolution of the Milky Way as his subjects while still a student in Leiden, and continued to work on them at Harvard. The dark, nebulous knots he suspected of being stellar birthplaces are now called Bok globules.

George Phillips Bond (May 20, 1825–February 17, 1865), the son of the observatory's founding director, William Cranch Bond, assisted in all his father's discoveries before taking over as director himself in 1859. He extended the early experiments in stellar photography and was the first American astronomer to win the gold medal of the Royal Astronomical Society.

Selina Cranch Bond (December 4, 1831–November 25, 1920), George's sister and the sixth child of William Cranch Bond, began working at the observatory as a teenager, was later hired as a computer, and continued lifelong in that profession.

William Cranch Bond (September 9, 1789–January 29, 1859), a successful chronometer maker before he became founding director of the observatory, established its time service, discovered (with son George) Saturn's inner ring and eighth satellite (Hyperion), and aided in taking the first-ever photograph of a star (Vega) in 1850.

Catherine Wolfe Bruce (January 22, 1816–March 13, 1900), a New York heiress who became an astronomy enthusiast in her later years, funded numerous research projects, journals, and instruments with the guidance of observatory director Edward Pickering, and also endowed a prestigious lifetime achievement award, the Bruce Medal.

Leon Campbell (January 20, 1881–May 10, 1951) traced light curves of variable stars and taught others the techniques. For many years he collected, collated, and published reports for the American Association of Variable Star Observers.

Annie Jump Cannon (December 11, 1863–April 13, 1941) classified the spectra of several hundred thousand stars for the nine-volume Henry Draper Catalogue and its Extension. Her system, with its "OBAFGKM" order of spectral classes, was internationally adopted in 1922 and remains in use today.

Seth Carlo Chandler (September 16, 1846–December 31, 1913), though only briefly on staff, maintained a close association with Harvard for thirty years. He worked as an actuary, pursuing variable stars in his spare time, and also wrote a code for sending astronomical announcements by telegraph.

Anna Palmer Draper (September 19, 1839–December 8, 1914) partnered with her husband, Dr. Henry Draper, in his telescope making and astrophotography. After his early death she assured his legacy by funding the continuation of his work at Harvard, which resulted in the classification system that bears his name.

Henry Draper, M.D. (March 7, 1837–November 20, 1882), followed his father, Dr. John William Draper, into medicine, astronomy, and photography. He was the first, in 1872, to capture the spectrum of a star on film, and followed that feat by imaging the faint stars in the Orion Nebula in 1882.

Sir Arthur Stanley Eddington (December 28, 1882–November 22, 1944), one of the first to appreciate Einstein's theories, traveled to Príncipe Island, off the west coast of Africa, for the 1919 total solar eclipse, and returned with proof of general relativity. The leader in efforts to describe the internal constitution of the stars, Eddington was knighted in 1930.

Priscilla Fairfield (later Bok) (April 14, 1896–November 19, 1975) taught astronomy at Smith College while measuring the widths of spectral lines on plates at Harvard. With her husband, Bart Bok, she coauthored *The Milky Way,* a book intended for nonspecialists; the couple revised and updated the 1941 original through a fourth edition in 1974.

Williamina Paton Stevens Fleming (May 15, 1857–May 21, 1911), the first woman to hold an official title at Harvard University, built a stellar classification scheme and also discovered ten novae and more than three hundred variable stars, all from her study of spectra on glass plates.

Caroline Furness (June 24, 1869–February 9, 1936) was the sixth individual and first woman to earn a doctorate in astronomy at Columbia University, in 1900. She taught the subject for twenty years at Vassar, her alma mater, where her students included Adelaide Ames and Harvia Wilson.

Boris Petrovič Gerasimovič (March 31, 1889–November 30, 1937), director of the Pulkovo Observatory in Russia, spent the years 1926 to 1929 at Harvard, and visited again in 1932. Accused at home of "servility" toward foreign science, he was executed in the Stalinist purges of that period.

Willard Peabody Gerrish (August 31, 1866–November 11, 1951), the observatory's resident mechanical genius, designed telescopes and clock drives that controlled the instruments' motion during long-exposure photography. The "Gerrish code" he devised replaced Seth Carlo Chandler's telegraphic announcement code in 1906.

George Ellery Hale (June 29, 1868–February 21, 1938), who spent a year as a young apprentice to Edward Pickering, later pursued solar spectroscopy. He established the *Astrophysical Journal,* and helped found both the American Astronomical Society and the International Astronomical Union, as well as the Yerkes, Mount Wilson, and Palomar observatories.

Margaret Harwood (March 19, 1885–February 6, 1979) became the first Astronomical Fellow of the Nantucket Maria Mitchell Association, and later the director of that association's observatory, a post she retained for forty-one years while studying asteroids of variable brightness.

Ejnar Hertzsprung (October 8, 1873–October 21, 1967), a native of Denmark long affiliated with the Leiden Observatory in the Netherlands, was first to seize on Henrietta Leavitt's period-luminosity relation to measure the distance to the Small Magellanic Cloud. He uncovered the existence of both giant and dwarf red stars, demonstrated the variability of Polaris (the North Star), and helped chart the general course of stellar evolution.

Lydia Swain Mitchell Hinchman (November 4, 1845–December 3, 1938) established the Nantucket Maria Mitchell Association in memory of her famous cousin, and furthered many of its activities, most notably the funding of fellowships for young women pursuing careers in astronomy.

Frank Scott Hogg (June 26, 1904–January 1, 1951) became Harvard's first Ph.D. in astronomy in 1928, following Cecilia Payne's 1925 Ph.D. from Radcliffe. As director of the David Dunlap Observatory near Toronto, he edited Canadian astronomy journals and studied the radial velocities of stars.

Edward Skinner King (May 31, 1861–September 10, 1931) supervised stellar photography at Harvard for four decades. He helped establish a uniform photometric scale, devised tests for the quality and consistency of photographic plates, and tried to discern the effects of interstellar dust on stellar magnitudes.

Henrietta Swan Leavitt (July 4, 1868–December 12, 1921) discovered thousands of variable stars. She was the first to note a relation between certain variables' peak brightness and the period over which their brightness varied—a relation that proved a valuable means for measuring distances across space.

Percival Lowell (March 13, 1855–November 12, 1916), brother of Harvard president Abbott Lawrence Lowell and poet Amy Lowell, built an observatory in Flagstaff, Arizona, where he studied Mars and pursued a ninth planet beyond Neptune.

Antonia Coetana de Paiva Pereira Maury (March 21, 1866–January 8, 1952), niece of Henry and Anna Draper, was the first female college graduate to work at the observatory. She discovered an early spectroscopic binary and devised a spectral classification scheme capable of distinguishing giant stars from dwarf stars.

Donald H. Menzel (April 11, 1901–December 14, 1976) was drawn to astronomy after seeing a total solar eclipse in 1918, and traveled to observe more eclipses than anyone before him. He first visited Harvard as Princeton professor Henry Norris Russell's graduate student in 1923, and succeeded Shapley as director in 1952.

Maria Mitchell (August 1, 1818–June 28, 1889) discovered a comet in 1847, the first American woman to do so. After family friend William Cranch Bond of Harvard announced her find, she won a gold medal from the king of Denmark. In 1865 Matthew Vassar invited her to become the first professor of astronomy at his new college for women, where she taught Antonia Maury.

John Stefanos Paraskevopoulos (June 20, 1889–March 15, 1951), known internationally as "Dr. Paras," guided the transfer of the Boyden Station from Arequipa, Peru, to South Africa, where he and his wife, Dorothy Block, added one hundred thousand plates to Harvard's collection.

Cecilia Helena Payne (later Gaposchkin) (May 10, 1900–December 7, 1979), among the first women to achieve a Ph.D. in astronomy—and the first person to earn one at Harvard—ascertained the temperatures of the different classes of stars and estimated the great abundance of hydrogen in them while doing the research for her dissertation.

Edward Bromfield Phillips (October 5, 1824–June 21, 1848), a Harvard classmate of George Bond, died a suicide, leaving the observatory $100,000. The Phillips Professorship and Phillips Library honor his memory.

Edward Charles Pickering (July 19, 1846–February 3, 1919), fourth and longest-presiding director of the observatory from 1877 to 1919, burnished its reputation while innovating in photometry, photography, and spectroscopy. He initiated the Draper Memorial spectral classification and the program of nightly all-sky photography. Elected president of the American Astronomical Society in 1905, he retained the office through repeated reelection until his death.

William Henry Pickering (February 15, 1858–January 16, 1938), younger brother of Edward, brought photography expertise from MIT to Harvard and served as first director of the Boyden Station at Arequipa. He focused his attention on observing the planets and their moons, discovering a satellite of Saturn, Phoebe, in 1899.

William Augustus Rogers (November 13, 1832–March 1, 1898) ascertained the positions of stars through a decade of observing the times at which each one crossed Harvard's local north-south meridian, and also performed two decades of calculations, in which he was assisted by his wife, née Rebecca Jane Titsworth.

Henry Norris Russell (October 25, 1877–February 18, 1957) of Princeton University, regarded as the dean of American astronomers during his lifetime, supervised the graduate work of Harlow Shapley and Donald Menzel. Industrious and influential, he studied stellar composition and evolution, the relationship of magnitude to classification, and the distinctions between giant and dwarf stars.

Helen B. Sawyer (later Hogg) (August 1, 1905–January 28, 1993) took up the study of globular clusters with Harlow Shapley. After completing her doctoral work at Harvard, she moved with her husband, Frank, to Canada, becoming the first woman to observe with large telescopes in British Columbia and Ontario. She popularized astronomy through her newspaper column and other writing.

Arthur Searle (October 21, 1837–October 23, 1920) served at the observatory for fifty-two years, including a period as acting director after Joseph Winlock died. He assisted Pickering in photometry, and taught astronomy classes at Radcliffe.

Harlow Shapley (November 2, 1885–October 20, 1972), the fifth director, from 1921 to 1952, added graduate education to the observatory's mission. Using Cepheid variables and the period-luminosity relation, he showed the Sun to be far from the center of the Milky Way, contrary to previous belief.

Martha Betz Shapley (August 3, 1890–January 24, 1981), "first lady" of the observatory, earned three degrees from the University of Missouri (B.S. in Education, 1910; A.B., 1911; M.A., 1913) before continuing Latin studies and German philology at Bryn Mawr. Her math ability enabled her to compute everything from the orbits of eclipsing binaries to the trajectories of ballistics for the U.S. Navy during World War II.

Winslow Upton (October 12, 1853–January 8, 1914) assisted only two years at Harvard before moving on to the U.S. Naval Observatory, the U.S. Signal Service, and Brown University, but he captured the 1877–1879 atmosphere in his spoof *The Observatory Pinafore.*

Arville D. Walker (August 2, 1883–August 5, 1963) joined the staff following her 1906 graduation from Radcliffe. In addition to work on variable stars and light curves of novae, she served as Harlow Shapley's secretary and a trusted adviser to the younger women at the observatory.

Margaret Walton (later Mayall) (January 27, 1902–December 6, 1995) cooperated closely with Annie Cannon on stellar classification, and completed the work on the Henry Draper Extension that had been left unfinished at Miss Cannon's death. She joined a special weapons group at MIT during World War II, and later served as Pickering Memorial Astronomer with the American Association of Variable Star Observers.

Oliver Clinton Wendell (May 7, 1845–November 5, 1912) assisted Edward Pickering for more than thirty years of photometry studies, paying particular attention to the changing light of variable stars.

Fred Lawrence Whipple (November 5, 1906–August 30, 2004), a comet expert, joined the Harvard Observatory in 1931 and became director of the Smithsonian Astrophysical Observatory in 1955. His contributions include the first tracking network for artificial satellites and the Whipple shield to protect spacecraft from damage by meteors.

Sarah Frances Whiting (August 23, 1847–September 12, 1927)

learned from Edward Pickering how to set up a practical physics laboratory, and established one at Wellesley College, where she taught and inspired Annie Jump Cannon.

Harvia Hastings Wilson (December 23, 1900–May 4, 1989), a 1923 Vassar alumna, delayed the start of her graduate studies till 1924 because of illness. At Harvard she studied the Magellanic Clouds, but returned to Vassar in 1925 as a physics instructor, then married accountant Hubert Stanley Russell in 1927.

Anna Winlock (September 15, 1857–January 3, 1904), eldest child of Joseph and Isabella Winlock, accompanied her father to Kentucky for the total solar eclipse of 1869, and started her thirty-year career as a Harvard computer shortly after his death.

Joseph Winlock (February 6, 1826–June 11, 1875) worked as a computer for—and later superintended—the *American Ephemeris and Nautical Almanac.* Appointed as third observatory director in 1866, he devoted himself to improving the existing instruments and acquiring new ones.

Ida Woods (September 16, 1870–unknown date 1940) began her career at the Harvard Observatory in 1893, following her graduation from Wellesley College. A charter member of the American Association of Variable Star Observers, she served as its first clerk. She aided Solon Bailey's research by discovering numerous variable stars in globular clusters, and also helped detect eight novae on photographic plates of the Milky Way.

Frances Woodworth Wright (April 30, 1897–July 30, 1989) came to Harvard in 1928, after teaching at Elmira College. During World War II she taught celestial navigation to U.S. Navy officers and also wrote a book on the subject. After earning her Radcliffe Ph.D. in astronomy in 1958 under Fred Whipple's supervision, she continued working until 1971.

Anne Sewell Young (January 2, 1871–August 15, 1961) earned a doctorate in astronomy from Columbia University and taught at Mount Holyoke for thirty-seven years. She took eight hundred students, including Helen Sawyer, from Smith and Mount Holyoke by train to see the January 1925 total solar eclipse in Windsor, Connecticut.

REMARKS

Preface

In its early years, the Harvard Observatory was often called the "Observatory at Cambridge." Its official naming in 1849 as the "Astronomical Observatory of Harvard College" set it apart from a meteorological observatory—and preserved the word *college,* even though Harvard, founded in 1636, had been recognized as a university since 1780.

The observatory's first home was at Dana House in Harvard Yard, but it moved in 1844 to Summerhouse Hill, which name gradually changed to Observatory Hill.

The first instruments of the Harvard College Observatory were those belonging to William Cranch Bond as his personal property.

CHAPTER ONE: Mrs. Draper's Intent

Mrs. Draper's full first name was Mary Anna, but she always signed herself as Anna Palmer Draper.

Dr. John William Draper, Henry's father, took the first photograph of the Moon, in 1839, and also, in 1840, one of the first photographic portraits made with sunlight. The subject was his sister Dorothy Catherine.

Scientists responded enthusiastically in 1877 to Dr. Henry Draper's detection of bright oxygen lines in the Sun's spectrum, but opposition arose within the year, especially among British observers such as Norman Lockyer. The main purpose of the Drapers' trip to England in 1879, when they visited William and Margaret Huggins, was to gain Henry an audience before the Royal Astronomical Society. After that presentation, he conducted additional research to defend his discovery, but died before announcing any further results. Controversy continued until 1896, when German physicists Carl Runge and Friedrich Paschen conclusively identified oxygen in the solar spectrum via dark Fraunhofer lines—not the bright lines that Draper had mistaken for proof.

Polaris later proved to be (slightly) variable. In 1911 the Danish astronomer Ejnar Hertzsprung detected a 0.14 magnitude change over not quite four days. Polaris is now known to be a multiple system made up of three component stars (one giant and two dwarfs).

CHAPTER TWO: What Miss Maury Saw

As Earth rotates daily and revolves annually, its north-south axis wobbles slowly over millennia, completing a full cycle every twenty-six thousand years.

As a result, the star that serves as a "polestar" changes over time. Our present North Star, Polaris, has no counterpart in the Southern Hemisphere.

Earth's wobble, called precession, changes the right ascension and declination of the stars by roughly one degree per century. Therefore nineteenth-century star catalogues gave stellar positions for a particular "epoch" date, such as 1875.0. Observations made in non-epoch years—1885, for example—were reduced (corrected by calculation) to 1880.0 or 1890.0.

Most of the naked-eye stars have individual names granted to them in the Middle Ages by Arabic astronomers, such as Altair for the brightest star in the Eagle, and Vega for the brightest in the Lyre. In the early seventeenth century the German astronomer Johann Bayer introduced a naming system using Greek letters, so that Vega was designated Alpha Lyrae, its next brightest constellation companion Beta Lyrae, and so on down the Greek alphabet as far as necessary. Although the Arabic star names persist in the West, Babylonian, Indian, Chinese, and other cultures' names have also adhered to the stars since antiquity.

John William Draper (1811–1882) met and married Antonia Coetana de Paiva Pereira Gardner (1814–1870) while visiting his relatives in England. The couple had six children: John Christopher (1835–1885), Henry (1837–1882), Virginia (1839–1885), Daniel (1841–1931), William (1845–1853), and Antonia (1849–1923). Dorothy Catherine Draper (1807–1901), whose self-sacrifice had facilitated her brother's education, also helped raise his children, as her sister-in-law was often ill. When Dorothy was thirty-two she had a serious suitor, but John William opposed the match, and she never married.

Antonia Maury's full name was Antonia Coetana de Paiva Pereira Maury. The de Paiva and Pereira families were Brazilian forebears of her grandmother Antonia Coetana de Paiva Pereira Gardner (Mrs. John William Draper).

Hermann Carl Vogel (1842–1907) of Germany independently discovered spectroscopic binaries at the same time as Edward Pickering. From his studies of spectra to gauge the motions of stars along the line of sight, Vogel showed that Algol and Spica each had an unseen companion.

Zeta Ursae Majoris, also known as Mizar, was split by telescope into two stars, Mizar A and B, photographed by George Bond in 1857. In 1889 Edward Pickering saw Mizar A itself as a pair—the first to be discovered by spectroscopy. Later, Mizar B also proved to be a binary pair.

CHAPTER THREE: Miss Bruce's Largesse

The $50,000 gift from Miss Bruce would equal well over $1 million in today's currency.

No portrait of Miss Bruce exists, as far as I can tell. Some searches turn up a handsome full-length portrait of a lady in a fur-trimmed yellow dress, but this is her cousin Catharine Lorillard Wolfe, also an heiress and a generous patroness of New York's Metropolitan Museum of Art.

Portraits of all the observatory directors hang on the walls at Harvard—all except for George Phillips Bond. Although he was a photography pioneer, he never had his own photograph taken or his likeness painted.

CHAPTER FOUR: Stella Nova

A nova, long thought to be a "new star," is now understood as the flaring of an ancient star in a binary system. The old star has exhausted its own fuel, but pulls in hydrogen from its companion. When enough hydrogen accumulates on the surface, an explosion of runaway fusion occurs to make the body suddenly visible. This can happen many times in a star's life. The objects observed by Tycho, Galileo, and Kepler are now classified as supernovae, or the end-stage catastrophic explosions of stars far more massive than our Sun. Since such an event destroys the star, the supernova phenomenon does not repeat.

Antonia Maury's sister Carlotta (1874–1938) attended Radcliffe College and Cornell and Columbia universities, and earned a doctorate in geology from Cornell in 1902. She traveled widely as a paleontologist, making numerous field trips to Brazil, Venezuela, South Africa, and several islands of the Caribbean. Another sister, Sarah, born in 1869, died in childhood. Their brother, John William Draper Maury (1871–1931), called Draper in his youth, attended Harvard and became a physician. He later dropped the name Maury.

CHAPTER FIVE: Bailey's Pictures from Peru

The founding of the Astronomical Society of the Pacific in February 1889 was a direct outcome of the previous month's eclipse. Staff members from the Lick Observatory and amateur astronomers and photographers in California enjoyed excellent viewing conditions and expedition results. They created an organization that remains a congenial mixture of professionals and amateurs, having grown from forty to six thousand members. The first woman admitted to membership was Rose O'Halloran, in 1892.

At first, the U.S. national professional organization of astronomers had no name. Hale very much wanted "astrophysics" to be part of the group's identity, and so it became the Astronomical and Astrophysical Society of America in 1899. As time passed, the name seemed unnecessarily unwieldy, especially as astrophysics came to dominate all of astronomy, and was changed to the American Astronomical Society in 1914.

Aristarchus, lacking good observational instruments, underestimated the distances to the Sun and Moon. The Sun is not twenty but fully four hundred times farther from Earth than the Moon.

CHAPTER SIX: Mrs. Fleming's Title

Edward Pickering also wrote a contribution for the "Chest of 1900," detailing his daily activities at the observatory as well as his leisure pursuits. In summers, he said, he liked to take long bicycle rides of twenty or thirty miles, two or three times a week. Since wheeling was his only exercise, he admitted that he suffered in the winter when he rode infrequently. On cloudy evenings Mrs.

Pickering often read to him from a novel. The two of them enjoyed playing chess together.

The anonymous gift of 1902 came from Henry H. Rogers of Standard Oil.

The asteroid Eros occasioned another worldwide observing campaign in 1975. The potato-shaped body is now known to rotate every five hours and has a varied composition that accounts for its changeable brightness.

CHAPTER SEVEN: Pickering's "Harem"

William H. Pickering claimed to have discovered a tenth satellite of Saturn in April 1905, and named it Themis, but it has yet to be confirmed.

CHAPTER EIGHT: Lingua Franca

Astronomer Friedrich Wilhelm Bessel announced the first successful distance measurement to a star, 61 Cygni, in 1838. He chose the star for its large proper motion, which suggested comparative nearness, then observed it along two different lines of sight. Just as a finger held in front of the face will appear to jump with respect to background objects when viewed first with one eye and then the other, a relatively close star will shift against the background stars when viewed at six-month intervals from opposite points in Earth's orbit (a baseline of two astronomical units). Bessel measured the star's angle of displacement, called parallax, and expressed the stellar distance in astronomical units, which translated to about ten light-years. It was a triumphant achievement. However, since stellar parallax angles are tiny, the method carried stellar distance measurements only so far—no more than a few hundred light-years from the Sun.

CHAPTER NINE: Miss Leavitt's Relationship

As Annie Jump Cannon, Antonia Maury, Henry Norris Russell, and others suspected, the various color categories in the Draper classification are indeed associated with specific stages in the lives of stars. Astronomers now know that only the most massive stars begin life in bright blue or white; as a consequence of burning so brightly, they burn out much faster than lower-mass stars such as the Sun. Our Sun, a G-type star, has existed for some five billion years and shines yellow, indicating a surface temperature of about six thousand degrees. In another few billion years, when it has converted most of its hydrogen to helium, the Sun will expand in diameter but cool at the surface, becoming a red giant M star. Other changes will ultimately render it a nonluminous "white dwarf."

CHAPTER TEN: The Pickering Fellows

The study of Cepheid variables affected areas of astrophysics beyond cosmic distances. Attempts by Arthur Stanley Eddington and others to explain what factors would make a star pulsate led, eventually, to an understanding of the structure, behavior, and longevity of stars in general.

chapter eleven: Shapley's "Kilo-Girl" Hours

The Nantucket Maria Mitchell Association chose Fiammetta Wilson as the 1920–1921 Edward C. Pickering Fellow, but she fell ill and died in July 1920 before learning of the award, which went instead to her colleague A. Grace Cook.

One or two capital letters at the start of a star name, such as SW Andromedae, identify that star as a variable. Stars that were named before anyone demonstrated their variability, such as Delta Cephei, retain their original names.

chapter twelve: Miss Payne's Thesis

The first woman to pursue advanced study in astronomy was Dorothea Klumpke, a native of San Francisco who conducted research on the rings of Saturn and earned a doctor of science degree at the University of Paris in 1893. Remaining in Europe, she worked for the French Bureau of Measurements and married English amateur astronomer Isaac Roberts.

The first woman to earn a Ph.D. in astronomy in the United States was Margaretta Palmer (1862–1924), at Yale, in 1894. A Vassar classmate of Antonia Maury's, she wrote her thesis on the orbit of Comet 1847 VI—the one discovered by her professor, Maria Mitchell. Dr. Palmer had been working at Yale as a computer before starting graduate studies, and stayed on the staff until her death.

The Henry Draper Extension, published in six installments between 1925 and 1936, added roughly another fifty thousand spectral classifications of faint stars to the quarter million given in the nine-volume Henry Draper Catalogue.

chapter thirteen: *The Observatory Pinafore*

The Harvard-Radcliffe Gilbert & Sullivan Players gave an abridged concert performance of *The Observatory Pinafore* at the American Academy of Arts & Sciences in Cambridge on October 26, 2000, as part of a banquet and centenary symposium honoring Cecilia Payne-Gaposchkin.

The so-called dark matter demonstrated by Robert Trumpler's research was interstellar dust. It should not be confused with the mysterious invisible entity given the same name by modern astronomers, who believe dark matter is what holds galaxies together.

chapter fourteen: Miss Cannon's Prize

After Harlow Shapley announced the wedding of Cecilia Payne and Sergei Gaposchkin, Miss Cannon made a note on the appropriate page in her diary. This particular diary was of the five-year variety, with room for only a paragraph-length entry per date, and in the allotted box she had already recorded how water pouring in through the back wall had flooded the cellar to a depth of several inches (though whether at her home or the observatory she did not specify). She had also mentioned teaching a class about the observatory's early history. Now, writing at a slant in the right-hand margin alongside these March 5 events, she added the news: "C.H.P. and S.G. married at Municipal Chapel, N.Y."

CHAPTER FIFTEEN: The Lifetimes of Stars

Beta Lyrae's orbital elements were computed completely and successfully for the first time in 2008. The component stars are so close together that the task remained impossible till this late date. Nearly one thousand Beta Lyrae–type variables are now known.

Many former members of the Harvard Observatory staff are interred in Cambridge, at the Mount Auburn Cemetery. It is a beautiful spot, as much an arboretum as a burial ground. Since the astronomers' graves are randomly distributed throughout the grounds, the cemetery office provides a map, with each relevant plot marked by a star. The Bond family members are reunited at Mount Auburn, as are the King family and the Bailey family, including Solon and Ruth's second son, Chester Romaña Bailey, only three months old when he died in August 1892. The Baileys were back in New England then, fresh from their first assignment in Peru, and the child's middle name honored good friends they had made in Arequipa.

Edward and Lizzie Pickering's paired headstones stand side by side. Hers identifies her as the wife of Edward Charles Pickering and daughter of Jared Sparks. His bears only one word in addition to his dates of birth and death, "Thanatopsis," the title of the poem about dying by William Cullen Bryant. The stone slab over Williamina Paton Fleming's grave also bears a solitary word, which describes her as she described herself in life: "Astronomer."

BIBLIOGRAPHY

Abir-Am, Pnina G., and Dorinda Outram, eds. *Uneasy Careers and Intimate Lives: Women in Science, 1789–1979.* New Brunswick, CT: Rutgers University Press, 1989.

Adams, Walter S. "The History of the International Astronomical Union." *Publications of the Astronomical Society of the Pacific* 61 (1949): 5–12.

Albers, Henry, ed. *Maria Mitchell: A Life in Journals and Letters.* Clinton Corners, NY: College Avenue Press, 2001.

Bailey, Solon I. "The Arequipa Station of the Harvard Observatory." *Popular Science Monthly* 64 (1904): 510–22.

_____. "Conditions in South Africa for Astronomical Observations." *Scientific Monthly* 21 (1925): 225–44.

_____. "Edward Charles Pickering, 1846–1919." *Astrophysical Journal* 50 (1919): 233–44.

_____. *The History and Work of Harvard Observatory, 1839 to 1927.* New York: McGraw Hill, 1931.

_____. "ω Centauri." *Astronomy and Astro-Physics* 12 (1893): 689–92.

_____. "The Study of Variable Stars." *Popular Science Monthly* 69 (1906): 175–85.

Baker, Daniel W. "History of the Harvard College Observatory During the Period 1840–1890" (pamphlet, reprinted from the six-article series in the Boston *Evening Traveller*). Cambridge, MA, 1890.

Bartusiak, Marcia, ed. *Archives of the Universe: 100 Discoveries That Transformed Our Understanding of the Cosmos.* New York: Pantheon, 2004.

_____. *The Day We Found the Universe.* New York: Pantheon, 2009.

Becker, Barbara J. *Unravelling Starlight: William and Margaret Huggins and the Rise of the New Astronomy.* Cambridge: Cambridge University Press, 2011.

Bergland, Renée. *Maria Mitchell and the Sexing of Science: An Astronomer Among the American Romantics.* Boston: Beacon, 2008.

Blaauw, Adriaan. *History of the IAU: The Birth and First Half-Century of the International Astronomical Union.* Dordrecht, Netherlands: Springer, 2012.

Boyd, Sylvia L. *Portrait of a Binary: The Lives of Cecilia Payne and Sergei Gaposchkin.* Rockland, ME: Penobscot Press, 2014.

Cahill, Maria J. "The Stars Belong to Everyone: Astronomer and Science Writer Helen Sawyer Hogg (1905–1993)." *Journal of the American Association of Variable Star Observers* 40 (2012): 31–43.

Chandler, S. C. "On the Observations of Variable Stars with the Meridian-Photometer of the Harvard College Observatory." *Astronomische Nachrichten* 134 (1894): 355–60.

Christianson, Gale E. *Edwin Hubble: Mariner of the Nebulae.* New York: Farrar, Straus and Giroux, 1995.

Clerke, Agnes M. *A Popular History of Astronomy During the Nineteenth Century.* Edinburgh: Adam & Charles Black; New York: Macmillan, 1887.

Coles, Peter. "Einstein, Eddington and the 1919 Eclipse." arXiv:astro-ph/0102462 (2001).

Collins, J. R. "The Royal Astronomical Society of Canada's Expedition to Observe the Total Eclipse of the Sun, August 31, 1932." *Journal of the Royal Astronomical Society of Canada* 26 (1932): 425–36.

Conway, Jill K. *The Female Experience in 18th and 19th Century America: A Guide to the History of American Women.* New York: Garland, 1982.

Des Jardins, Julie. *The Madame Curie Complex: The Hidden History of Women in Science.* New York: Feminist Press, 2010.

DeVorkin, David H. "Community and Spectral Classification in Astrophysics: The Acceptance of E. C. Pickering's System in 1910." *Isis* 72 (1981): 29–49.

_____. *Henry Norris Russell: Dean of American Astronomers.* Princeton, NJ: Princeton University Press, 2000.

DeVorkin, David H., and Ralph Kenat. "Quantum Physics and the Stars (III): Henry Norris Russell and the Search for a Rational Theory of Stellar Spectra." *Journal for the History of Astronomy* 21 (1990): 157–86.

Dick, Steven J. *Sky and Ocean Joined: The U.S. Naval Observatory, 1830–2000.* Cambridge: Cambridge University Press, 2003.

Dobson, Andrea K., and Katherine Bracher. "A Historical Introduction to Women in Astronomy." *Mercury* 21 (1992): 4–15.

Draper, Henry. *On the Construction of a Silvered Glass Telescope, Fifteen and a Half Inches in Aperture, and Its Uses in Celestial Photography.* Washington, DC: Smithsonian Institution, 1864.

_____. "Researches upon the Photography of Planetary and Stellar Spectra." *Proceedings of the American Academy of Arts and Sciences* 19 (1884): 231–61.

Fernie, J. D. "The Historical Quest for the Nature of the Spiral Nebulae." *Publications of the Astronomical Society of the Pacific* 82 (1970): 1189–230.

_____. "The Period-Luminosity Relation: A Historical Review." *Publications of the Astronomical Society of the Pacific* 81 (1969): 707–31.

Frost, Edwin B. "A Desideratum in Spectrology." *Astrophysical Journal* 20 (1904): 342–45.

Gingerich, Owen. "How Shapley Came to Harvard, or, Snatching the Prize from the Jaws of Debate." *Journal for the History of Astronomy* 19 (1988): 201–7.

Gingrich, C. H. "The Fifth Conference of the International Union for Co-operation in Solar Research." *Popular Astronomy* 21 (1913): 457–68.

Glass, I. S. *Revolutionaries of the Cosmos: The Astro-Physicists.* Oxford: Oxford University Press, 2006.

Grindlay, Jonathan E., and A. G. Davis Philip, eds. *The Harlow Shapley Symposium on Globular Cluster Systems in Galaxies.* Dordrecht, Netherlands: Kluwer, 1988.

Hale, George Ellery. *The New Heavens.* New York: Charles Scribner's Sons, 1922.

Hall, G. Harper. "The Total Eclipse of 1932." *Journal of the Royal Astronomical Society of Canada* 26: 337–44.

Haramundanis, Katherine, ed. *Cecilia Payne-Gaposchkin: An Autobiography and Other Recollections.* 2nd ed. Cambridge: Cambridge University Press, 1996.

Hearnshaw, John B. *The Analysis of Starlight: Two Centuries of Astronomical Spectroscopy.* 2nd ed. New York: Cambridge University Press, 2014.

_____. *The Measurement of Starlight: Two Centuries of Astronomical Photometry.* Cambridge: Cambridge University Press, 1996.

Henden, Arne A., and Ronald H. Kaitchuck. *Astronomical Photometry: A Text and Handbook for the Advanced Amateur and Professional Astronomer.* New York: Van Nostrand Reinhold, 1982.

Hirshfeld, Alan W. *Parallax: The Race to Measure the Cosmos.* New York: W. H. Freeman, 2001.

_____. *Starlight Detectives: How Astronomers, Inventors, and Eccentrics Discovered the Modern Universe.* New York: Bellevue Literary Press, 2014.

Hoar, Roger Sherman. "The Pickering Polaris Attachment." *Journal of the United States Artillery* 50 (1919): 230–36.

Hoffleit, Dorrit. "E. C. Pickering in the History of Variable Star Astronomy." *Journal of the American Association of Variable Star Observers* 1 (1972): 3–8.

_____. *Maria Mitchell's Famous Students.* Cambridge, MA: American Association of Variable Star Observers, 1983.

_____. *Misfortunes as Blessings in Disguise.* Cambridge, MA: American Association of Variable Star Observers, 2002.

_____. *Women in the History of Variable Star Astronomy.* Cambridge, MA: American Association of Variable Star Observers, 1993.

Hoskin, M. A. "The 'Great Debate': What Really Happened." *Journal for the History of Astronomy* 7 (1976): 169–82.

_____. *Stellar Astronomy: Historical Studies.* Chalfont St. Giles, Bucks, UK: Science History Publications, 1982.

Hughes, Patrick. *A Century of Weather Service: A History of the Birth and Growth of the National Weather Service, 1870–1970.* New York: Gordon and Breach, 1970.

James, Edward T., Janet Wilson James, and Paul S. Boyer, eds. *Notable American Women, 1607–1950: A Biographical Dictionary.* 3 vols. Cambridge, MA: Belknap Press of Harvard University Press, 1971.

Johnson, George. *Miss Leavitt's Stars: The Untold Story of the Woman Who Discovered How to Measure the Universe.* New York: Norton, 2005.

Jones, Bessie Zaban, and Lyle Gifford Boyd. *The Harvard College Observatory: The First Four Directorships, 1839–1919.* Cambridge, MA: Belknap Press of Harvard University Press, 1971.

Kass-Simon, G., and Patricia Farnes, eds. *Women of Science: Righting the Record.* Bloomington: Indiana University Press, 1990.

Kennefick, Daniel. "Testing Relativity from the 1919 Eclipse: A Question of Bias." *Physics Today,* March 2009, 37–42.

Lafortune, Keith R. "Women at the Harvard College Observatory, 1877–1919: 'Women's Work,' the 'New' Sociality of Astronomy, and Scientific Labor." Master's thesis, University of Notre Dame, 2001.

Langley, Samuel P. *The New Astronomy.* Boston: Ticknor, 1888.

Lankford, John. *American Astronomy: Community, Careers and Power, 1859–1940.* Chicago: University of Chicago Press, 1997.

Levy, David H. *The Man Who Sold the Milky Way: A Biography of Bart Bok.* Tucson: University of Arizona Press, 1993.

Lockyer, J. Norman. *Elementary Lessons in Astronomy.* London: Macmillan, 1889.

Mack, Pamela Etter. "Women in Astronomy in the United States 1875–1920." Bachelor's thesis, Harvard University, April 1977.

McLaughlin, Dean B. "The Fifty-third Meeting of the American Astronomical Society." *Popular Astronomy* 43 (1935): 75–78.

Mozans, H. J. (anagrammatized pen name of the Reverend John A. Zahm). *Woman in Science.* New York: Appleton, 1913.

Newcomb, Simon. "The Place of Astronomy Among the Sciences." *Sidereal Messenger* 7 (1888): 65–73.

North, John. *The Norton History of Astronomy and Cosmology.* New York: Norton, 1994.

Ogilvie, Marilyn Bailey. *Women in Science: Antiquity Through the Nineteenth Century; A Biographical Dictionary with Annotated Bibliography.* Cambridge, MA: MIT Press, 1990.

Pasachoff, Jay M., and Terry-Ann Suer. "The Origin and Diffusion of the H and K Notation." *Journal of Astronomical History and Heritage* 13 (2010): 120–26.

Payne, Cecilia Helena. *Stellar Atmospheres: A Contribution to the Observational Study of High Temperature in the Reversing Layers of Stars.* Cambridge, MA: Harvard College Observatory, 1925.

Payne-Gaposchkin, Cecilia. "The Dyer's Hand: An Autobiography." 1979. Published posthumously in *Cecilia Payne-Gaposchkin: An Autobiography and Other Recollections,* 2nd ed., edited by Katherine Haramundanis, 69–238. Cambridge: Cambridge University Press, 1996.

_____. *Introduction to Astronomy.* New York: Prentice-Hall, 1954.

_____. *Stars in the Making.* Cambridge, MA: Harvard University Press, 1952.

Peed, Dorothy Myers. *America Is People and Ideas: Library Researching for the Space Age.* New York: Exposition, 1966.

Philip, A. G. Davis, and Rebecca A. Koopmann, eds. *The Starry Universe: The Cecilia Payne-Gaposchkin Centenary.* Proceedings of a symposium held at the Harvard-Smithsonian Center for Astrophysics, Cambridge, Massachusetts, October 26–27, 2000. Schenectady, NY: L. Davis, 2001.

Pickering, Edward C. "A New Star in Norma." *Astronomy and Astro-Physics* 13 (1893): 40–41.

_____. "On the Spectrum of Zeta Ursae Majoris." *American Journal of Science,* 3rd ser., 39 (1890): 46–47.

_____. *Statement of Work Done at the Harvard College Observatory During the Years 1877–1882.* Cambridge, MA: John Wilson & Son University Press, 1882.

Pickering, William H. "Mars." *Astronomy and Astro-Physics* 11 (1892): 668–75.

Plaskett, J. S. "The Astronomical and Astrophysical Society of America." *Journal of the Royal Astronomical Society of Canada* 4 (1910): 373–78.

_____. "The Solar Union." *Journal of the Royal Astronomical Society of Canada* 7 (1913): 420–37.

Plotkin, Harold. "Edward Charles Pickering." *Journal for the History of Astronomy* 21 (1990): 47–58.

_____. "Edward Charles Pickering's Diary of a Trip to Pasadena to Attend Meeting of Solar Union, August 1910." *Southern California Quarterly* 60 (1978): 29–44.

———. "Edward C. Pickering and the Endowment of Scientific Research in America, 1877–1918." *Isis* 69 (1978): 44–57.

———. "Edward C. Pickering, the Henry Draper Memorial, and the Beginnings of Astrophysics in America." *Annals of Science* 35 (1978): 365–77.

———. "Harvard College Observatory's Boyden Station in Peru: Origin and Formative Years, 1879–1898." In *Mundialización de la ciencia y cultura nacional: Actas del Congreso Internacional "Ciencia, Descubrimento y Mondo Colonial,"* edited by A. Lafuente, A. Elena, and M. L. Ortega, 689–705. Madrid: Doce Calles, 1993.

———. "Henry Draper, the Discovery of Oxygen in the Sun, and the Dilemma of Interpreting the Solar Spectrum." *Journal for the History of Astronomy* 8 (1977): 44–51.

———. "William H. Pickering in Jamaica: The Founding of Woodlawn and Studies of Mars." *Journal for the History of Astronomy* 24 (1993): 101–22.

Putnam, William Lowell. *The Explorers of Mars Hill: A Centennial History of Lowell Observatory, 1894–1994.* West Kennebunk, ME: Phoenix, 1994.

Rossiter, Margaret W. *Women Scientists in America: Struggles and Strategies to 1940.* Baltimore: Johns Hopkins University Press, 1982.

Rubin, Vera. *Bright Galaxies, Dark Matters.* New York: Springer-Verlag, 1996.

Sadler, Philip M. "William Pickering's Search for a Planet Beyond Neptune." *Journal for the History of Astronomy* 21 (1990): 59–64.

Schechner, Sara J., and David H. Sliski. "The Scientific and Historical Value of Annotations on Astronomical Photographic Plates." *Journal for the History of Astronomy* 47 (2016): 3–29.

Schlesinger, Frank. "The Astronomical and Astrophysical Society of America." *Science* 32 (1910): 874–87.

Shapley, Harlow. "On the Nature and Cause of Cepheid Variation." *Astrophysical Journal* 40 (1914): 448–65.

———. *Through Rugged Ways to the Stars.* New York: Charles Scribner's Sons, 1969.

Shapley, Harlow, and Cecilia H. Payne, eds. *The Universe of Stars.* Cambridge, MA: Harvard Observatory, 1929.

Smith, Horace A. "Bailey, Shapley, and Variable Stars in Globular Clusters." *Journal for the History of Astronomy* 31 (2000): 185–201.

Smith, Robert W. *The Expanding Universe: Astronomy's "Great Debate," 1900–1931.* Cambridge: Cambridge University Press, 1982.

Spradley, Joseph L. "Women and the Stars." *Physics Teacher* 28 (Sept. 1990): 372–77.

Stanley, Matthew. "The Development of Early Pulsation Theory, or, How Cepheids Are Like Steam Engines." *Journal of the American Association of Variable Star Observers* 40 (2012): 100–108.

Strauss, David. *Percival Lowell: The Culture and Science of a Boston Brahmin.* Cambridge, MA: Harvard University Press, 2001.

Tenn, Joseph S. "A Brief History of the Bruce Medal of the A.S.P." *Mercury* 15 (1986): 103–11.

Wayman, Patrick. "Cecilia Payne-Gaposchkin: Astronomer Extraordinaire." *Astronomy & Geophysics* 43 (2002): 1.27–1.29.

Williams, Thomas R., and Michael Saladyga. *Advancing Variable Star Astronomy.* Cambridge: Cambridge University Press, 2011.

Wilson, H. C. "The Fourth Conference of the International Union for Co-operation in Solar Research." *Popular Astronomy* 18 (1910): 489–503.

Young, Charles A. "The Great Comet of 1882." *Popular Science Monthly* 22 (Jan. 1883): 289–300.

Zerwick, Chloe. *A Short History of Glass.* New York: Abrams, 1990.

INDEX